Multivariate Statistica

A PRIMER

A journey of a thousand miles begins with a single step

Lao Tsu

———————— TO JEAN ————————

Multivariate Statistical Methods
A PRIMER

Bryan F.J. Manly

Department of Mathematics and Statistics
University of Otago
New Zealand

LONDON   NEW YORK
Chapman and Hall

First published in 1986 by
Chapman and Hall Ltd
11 New Fetter Lane, London EC4P 4EE
Published in the USA by
Chapman and Hall
29 West 35th Street, New York NY 10001
Reprinted 1988

©1986 Bryan F.J. Manly

Printed in Great Britain by J.W. Arrowsmith Ltd, Bristol

ISBN 0 412 28610 6 (hardback)
      0 412 28620 3 (paperback)

British Library Cataloguing in Publication Data

Manly, Bryan F.J.
Multivariate statistical methods: A primer
    1. Multivariate analysis
    I. Title
    519.5'35   QA278

    ISBN 0-412-28610-6
    ISBN 0-412-28620-3 Pbk

Library of Congress Cataloging in Publication Data

Manly, Bryan F.J., 1944–
    Multivariate statistical methods: A Primer

    Includes bibliographies and indexes.
    1. Multivariate analysis.
    I. Title.
QA278.M35 1986   519.5'35   86–9537
ISBN 0-412-28610-6
ISBN 0-412-28620-3 (pbk.)

Contents

v

Contents

Contents

Preface

The purpose of this book is to introduce multivariate statistical methods to non-mathematicians. It is not intended to be particularly comprehensive. Rather, the intention is to keep details to a minimum while still conveying a good idea of what can be done. In other words, it is a book to 'get you going' in a particular area of statistical methods.

It is assumed that readers have a working knowledge of elementary statistics, particularly tests of significance using the normal, t, chi-square and F distributions, analysis of variance, and linear regression. The material covered in a standard first-year university service course in statistics should be quite adequate, together with a reasonable facility with ordinary algebra. Also, understanding multivariate analysis requires some use of matrix algebra. However, the amount needed is quite small if one is prepared to accept certain details on faith. Anyone who masters the material in Chapter 2 will have the required basic minimum level of matrix competency.

One of the reasons why multivariate methods are being used so much these days is obviously the availability of statistical packages to do the calculations. For most people access to a package will be a necessity for using the methods. However, statistical packages are not stressed particularly in this book for two reasons. First, there are so many packages available that it is not realistic to write about them all, or to concentrate on any particular one. Second, the calculations for many analyses are relatively straightforward, and can be programmed on a microcomputer fairly easily. The approach adopted in this book is therefore in most cases to state what calculations have to be done and to give some idea of how these can be programmed. Readers who want to do their own analyses can choose a convenient package and read the manual to find out how to use it.

To some extent the chapters of this book can be read independently of each other. The first four are preliminary reading in that they are largely concerned with general aspects of handling multivariate data rather than with specific techniques. Chapter 1 introduces some examples that are used in subsequent chapters and briefly describes the six multivariate methods of analysis that this book is primarily concerned with. As mentioned above, Chapter 2 provides the minimum level of matrix competency required for understanding the remainder of the book. Chapter 3 is about tests of significance and is not crucial as far as understanding the following chapters is concerned. Chapter 4 is about measuring distances with multivariate data. At least the first four sections of this chapter should be read before Chapters 7, 8 and 10.

Chapters 5 to 10 cover what I consider to be the most important multivariate techniques of data analysis. Of these, Chapters 5 and 6 form a natural pair to be read together. However, Chapters 7 to 10 can be read singly and still (I hope) make sense.

Finally, in Chapter 11, I have attempted to sum up what has been covered and make some general comments on good practices with the analysis of multivariate data. The Appendix contains three example sets of data for readers to analyse by themselves.

I am indebted to many people for their comments on the various draft versions of this book. Earl Bardsley read early versions of several of the chapters. Anonymous reviewers read all or parts of the work. John Harraway read through the final version. Their comments have led to numerous improvements. However, I take all responsibility for any errors.

Mary–Jane Campbell cheerfully typed and retyped the manuscript as I made changes. I am most grateful to her.

B.F.J. Manly
Dunedin, November 1985

CHAPTER ONE

The material of
multivariate analysis

1.1 Examples of multivariate data

The statistical methods that are described in elementary texts are mostly univariate methods because they are only concerned with analysing variation in a single random variable. This is even true of multiple regression because this technique involves trying to account for variation in one dependent variable. On the other hand, the whole point of a multivariate analysis is to consider several related random variables simultaneously, each one being considered equally important at the start of the analysis. The potential value of this more general approach is perhaps best seen by considering a few examples.

Example 1.1 Storm survival of sparrows

After a severe storm on 1 February 1898, a number of moribund sparrows were taken to the biological laboratory at Brown University, Rhode Island. Subsequently about half of the birds died and Hermon Bumpus saw this as an opportunity to study the effect of natural selection on the birds. He took eight morphological measurements on each bird and also weighed them. The results for five of the variables are shown in Table 1.1, for females only.

When Bumpus collected his data in 1898 his main interest was in the light that it would throw on Darwin's theory of natural selection. He concluded from studying the data that 'the birds which perished, perished not through accident, but because they were physically disqualified, and that the birds which survived, survived because they possessed certain physical characters.' To be specific, the survivors 'are shorter and weigh less...have longer wing bones, longer legs, longer sternums and greater brain capacity' than the non-survivors.

1

Table 1.1 Body measurements of female sparrows (X_1 = total length, X_2 = alar extent, X_3 = length of beak and head, X_4 = length of humerus, X_5 = length of keel of sternum; all in mm). Birds 1 to 21 survived, while the remainder died.

Bird	X_1	X_2	X_3	X_4	X_5
1	156	245	31.6	18.5	20.5
2	154	240	30.4	17.9	19.6
3	153	240	31.0	18.4	20.6
4	153	236	30.9	17.7	20.2
5	155	243	31.5	18.6	20.3
6	163	247	32.0	19.0	20.9
7	157	238	30.9	18.4	20.2
8	155	239	32.8	18.6	21.2
9	164	248	32.7	19.1	21.1
10	158	238	31.0	18.8	22.0
11	158	240	31.3	18.6	22.0
12	160	244	31.1	18.6	20.5
13	161	246	32.3	19.3	21.8
14	157	245	32.0	19.1	20.0
15	157	235	31.5	18.1	19.8
16	156	237	30.9	18.0	20.3
17	158	244	31.4	18.5	21.6
18	153	238	30.5	18.2	20.9
19	155	236	30.3	18.5	20.1
20	163	246	32.5	18.6	21.9
21	159	236	31.5	18.0	21.5
22	155	240	31.4	18.0	20.7
23	156	240	31.5	18.2	20.6
24	160	242	32.6	18.8	21.7
25	152	232	30.3	17.2	19.8
26	160	250	31.7	18.8	22.5
27	155	237	31.0	18.5	20.0
28	157	245	32.2	19.5	21.4
29	165	245	33.1	19.8	22.7
30	153	231	30.1	17.3	19.8
31	162	239	30.3	18.0	23.1
32	162	243	31.6	18.8	21.3
33	159	245	31.8	18.5	21.7
34	159	247	30.9	18.1	19.0
35	155	243	30.9	18.5	21.3
36	162	252	31.9	19.1	22.2
37	152	230	30.4	17.3	18.6
38	159	242	30.8	18.2	20.5
39	155	238	31.2	17.9	19.3
40	163	249	33.4	19.5	22.8

Table 1.1 (*Contd.*)

Bird	X_1	X_2	X_3	X_4	X_5
41	163	242	31.0	18.1	20.7
42	156	237	31.7	18.2	20.3
43	159	238	31.5	18.4	20.3
44	161	245	32.1	19.1	20.8
45	155	235	30.7	17.7	19.6
46	162	247	31.9	19.1	20.4
47	153	237	30.6	18.6	20.4
48	162	245	32.5	18.5	21.1
49	164	248	32.3	18.8	20.9

Data source: Bumpus (1898).

He also concluded that 'the process of selective elimination is most severe with extremely variable individuals, no matter in which direction the variations may occur. It is quite as dangerous to be conspicuously above a certain standard of organic excellence as it is to be conspicuously below the standard.' This last statement is saying that stabilizing selection occurred, so that individuals with measurements close to the average survived better than individuals with measurements rather different from the average.

Of course, the development of multivariate statistical methods had hardly begun in 1898 when Bumpus was writing. The correlation coefficient as a measure of the relationships between two variables was introduced by Francis Galton in 1877. However, it was another 56 years before Hotelling described a practical method for carrying out a principal component analysis, which is one of the simplest multivariate analyses that can be applied to Bumpus's data. In fact Bumpus did not even calculate standard deviations. Nevertheless, his methods of analysis were sensible. Many authors have reanalysed his data and, in general, have confirmed his conclusions.

Taking the data as an example for illustrating multivariate techniques, several interesting questions spring to mind. In particular:

1. How are the different measurements related? For example, does a large value for one variable tend to occur with large values for the other variables?
2. Do the survivors and non-survivors have significant differences for the mean values of the variables?

Table 1.2 Measurements on male Egyptian skulls from various epochs (X_1 = maximum breadth, X_2 = basibregmatic height, X_3 = basialveolar length, X_4 = nasal height; all in mm, as shown on Fig. 1.1)

Skull	Early predynastic				Late predynastic				12th & 13th dynasties				Ptolemaic period				Roman period			
	X_1	X_2	X_3	X_4	X_1	X_2	X_3	X_4	X_1	X_2	X_3	X_4	X_1	X_2	X_3	X_4	X_1	X_2	X_3	X_4
1	131	138	89	49	124	138	101	48	137	141	96	52	137	134	107	54	137	123	91	50
2	125	131	92	48	133	134	97	48	129	133	93	47	141	128	95	53	136	131	95	49
3	131	132	99	50	138	134	98	45	132	138	87	48	141	130	87	49	128	126	91	57
4	119	132	96	44	148	129	104	51	130	134	106	50	135	131	99	51	130	134	92	52
5	136	143	100	54	126	124	95	45	134	134	96	45	133	120	91	46	138	127	86	47
6	138	137	89	56	135	136	98	52	140	133	98	50	131	135	90	50	126	138	101	52
7	139	130	108	48	132	145	100	54	138	138	95	47	140	137	94	60	136	138	97	58
8	125	136	93	48	133	130	102	48	136	145	99	55	139	130	90	48	126	126	92	45
9	131	134	102	51	131	134	96	50	136	131	92	46	140	134	90	51	132	132	99	55
10	134	134	99	51	133	125	94	46	126	136	95	56	138	140	100	52	139	135	92	54
11	129	138	95	50	133	136	103	53	137	129	100	53	132	133	90	53	143	120	95	51
12	134	121	95	53	131	139	98	51	137	139	97	50	134	134	97	54	141	136	101	54
13	126	129	109	51	131	136	99	56	136	126	101	50	135	135	99	50	135	135	95	56
14	132	136	100	50	138	134	98	49	137	133	90	49	133	136	95	52	137	134	93	53
15	141	140	100	51	130	136	104	53	129	142	104	47	136	130	99	55	142	135	96	52

16	131	134	97	54	131	128	98	45	135	138	102	55	134	137	93	52	139	134	95	47
17	135	137	103	50	138	129	107	53	129	135	92	50	131	141	99	55	138	125	99	51
18	132	133	93	53	123	131	101	51	134	125	90	60	129	135	95	47	137	135	96	54
19	139	136	96	50	130	129	105	47	138	134	96	51	136	128	93	54	133	125	92	50
20	132	131	101	49	134	130	93	54	136	135	94	53	131	125	88	48	145	129	89	47
21	126	133	102	51	137	136	106	49	132	130	91	52	139	130	94	53	138	136	92	46
22	135	135	103	47	126	131	100	48	133	131	100	50	144	124	86	50	131	129	97	44
23	134	124	93	53	135	136	97	52	138	137	94	51	141	131	97	53	143	126	88	54
24	128	134	103	50	129	126	91	50	130	127	99	45	130	131	98	53	134	124	91	55
25	130	130	104	49	134	139	101	49	136	133	91	49	133	128	92	51	132	127	97	52
26	138	135	100	55	131	134	90	53	134	123	95	52	138	126	97	54	137	125	85	57
27	128	132	93	53	132	130	104	50	136	137	101	54	131	142	95	53	129	128	81	52
28	127	129	106	48	130	132	93	52	133	131	96	49	136	138	94	55	140	135	103	48
29	131	136	114	54	135	132	98	54	138	133	100	55	132	136	92	52	147	129	87	48
30	124	138	101	46	130	128	101	51	138	133	91	46	135	130	100	51	136	133	97	51

Data source: Thomson and Randall-Maciver (1905).

3. Do survivors and non-survivors show the same amount of variation in measurements?
4. If the survivors and non-survivors differ with regard to their distributions for the variables, is it possible to construct some function of these variables $f(X_1, X_2, X_3, X_4, X_5)$ which separates the two groups? It would be convenient if this function tended to be large for survivors and small for non-survivors since it would then be an index of Darwinian fitness.

Example 1.2 Egyptian skulls

For a second example, consider the data shown in Table 1.2 for measurements made on male Egyptian skulls from the area of Thebes. There are five samples of 30 skulls from each of the early predynastic period (*circa* 4000 BC), the late predynastic period (*circa* 3300 BC), the 12th and 13th dynasties (*circa* 1850 BC), the Ptolemaic period (*circa* 200 BC), and the Roman period (*circa* AD 150). Four measurements are available on each skull, these being as shown in Fig. 1.1.

In this case it is interesting to consider the questions:

1. How are the four measurements related?
2. Are there significant differences in the sample means for the variables and, if so, do these differences reflect gradual changes with time?
3. Are there significant differences in the sample standard deviations

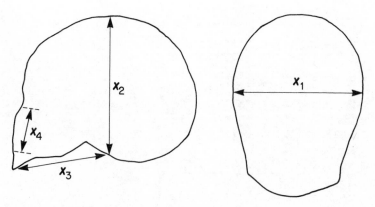

Figure 1.1 Measurements on Egyptian skulls.

for the variables and, if so, do the differences reflect gradual changes with time?

4. Is it possible to construct a function $f(X_1, X_2, X_3, X_4)$ of the four variables that in some sense captures most of the sample differences?

These questions are, of course, rather similar to the one suggested with Example 1.1.

As will be seen later, there are differences between the five samples that can be explained partly as time trends. It must be said, however, that the reasons for the changes are not known. Migration into the population was probably the most important factor.

Example 1.3 Distribution of a butterfly

A study of 16 colonies of the butterfly *Euphydryas editha* in California and Oregon produced the data shown in Table 1.3. Here there are two types of variable: environmental and distributional. The environmental variables are altitude, rainfall, and minimum and maximum temperatures. The distribution variables are gene frequencies for phosphoglucose-isomerase (Pgi) as determined by the technique of electrophoresis. For the present purposes there is no need to go into the details of how the gene frequencies were determined. (Strictly speaking these are not gene frequencies anyway.) It is sufficient to say that the frequencies describe the genetic distribution of *E. editha* to some extent. Figure 1.2 shows the geographical distribution of the colonies.

In this example questions that can be asked are:

1. Are the Pgi frequencies similar for colonies that are close in space?
2. To what extent are the Pgi frequencies related to the environmental variables?

These questions are important when it comes to trying to decide how Pgi frequencies are determined. If frequencies are largely determined by present and past migration then gene frequencies should be similar for adjacent colonies but may show no relationships with environmental variables. On the other hand, if it is the environment that is most important then the Pgi frequencies should be related to the environmental variables, but colonies that are close in space will have different frequencies if the environments are different. Of course,

Table 1.3 Environmental variables and phosphoglucose-isomerase (Pgi) gene frequencies for colonies of the butterfly *Euphydryas editha* in California and Oregon.

Colony	Altitude (feet)	Annual precipitation (inches)	Annual maximum temperature (°F)	Annual minimum temperature (°F)	Pgi mobility gene frequencies (%)					
					0.40	0.60	0.80	1.00	1.16	1.30
SS	500	43	98	17	0	3	22	57	17	1
SB	800	20	92	32	0	16	20	38	13	13
WSB	570	28	98	26	0	6	28	46	17	3
JRC	550	28	98	26	0	4	19	47	27	3
JRH	550	28	98	26	0	1	8	50	35	6
SJ	380	15	99	28	0	2	19	44	32	3
CR	930	21	99	28	0	0	15	50	27	8
UO	650	10	101	27	10	21	40	25	4	0
LO	600	10	101	27	14	26	32	28	0	0
DP	1500	19	99	23	0	1	6	80	12	1
PZ	1750	22	101	27	1	4	34	33	22	6
MC	2000	58	100	18	0	7	14	66	13	0
IF	2500	34	102	16	0	9	15	47	21	8
AF	2000	21	105	20	3	7	17	32	27	14
GH	7850	42	84	5	0	5	7	84	4	0
GL	10500	50	81	−12	0	3	1	92	4	0

Data source: McKechnie *et al.* (1975). Environmental variables have been rounded to integers for simplicity. The original data were for 21 colonies. For the present example, five colonies with small samples for the estimation of gene frequencies have been excluded so as to make all of the estimates about equally reliable.

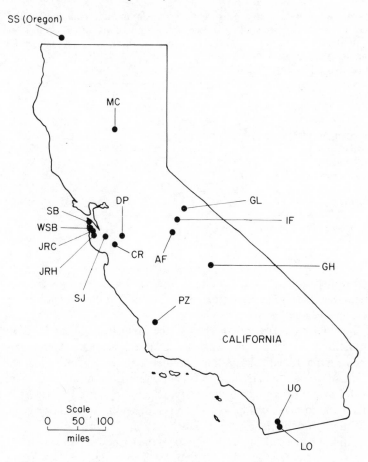

Figure 1.2 Colonies of *Euphydryas editha* in California and Oregon.

colonies that are close in space will tend to have similar environments.
It may therefore be difficult to reach a clear conclusion.

Example 1.4 Prehistoric dogs from Thailand

Excavations of prehistoric sites in northeast Thailand have produced
a series of canid (dog) bones covering a period from about 3500 BC to
the present. The origin of the prehistoric dog is not certain. It could
descend from the golden jackal (*Canis aureus*) or the wolf. However,

Table 1.4 Mean mandible measurements (mm) for modern Thai dogs, golden jackals, wolves, cuons, dingos and prehistoric dogs (X_1 = breadth of mandible, X_2 = height of mandible below 1st molar, X_3 = length of 1st molar, X_4 = breadth of 1st molar, X_5 = length from 1st to 3rd molars inclusive, X_6 = length from 1st to 4th premolars inclusive).

Group	X_1	X_2	X_3	X_4	X_5	X_6
Modern dog	9.7	21.0	19.4	7.7	32.0	36.5
Golden jackal	8.1	16.7	18.3	7.0	30.3	32.9
Chinese wolf	13.5	27.3	26.8	10.6	41.9	48.1
Indian wolf	11.5	24.3	24.5	9.3	40.0	44.6
Cuon	10.7	23.5	21.4	8.5	28.8	37.6
Dingo	9.6	22.6	21.1	8.3	34.4	43.1
Prehistoric dog	10.3	22.1	19.1	8.1	32.3	35.0

Data source: Higham *et al.* (1980).

the wolf is not native to Thailand, the nearest indigenous sources being western China (*Canis lupus chanco*) or the Indian subcontinent (*Canis lupus pallipes*).

In order to clarify the ancestry of the prehistoric dogs, mandible measurements were made on the available specimens. These were then compared with similar measurements on the golden jackal, the Chinese wolf and the Indian wolf. The comparisons were made more useful by considering also the dingo, which may have its origins in India, the cuon (*Cuon alpinus*) which is indigenous to southeast Asia, and modern village dogs from Thailand.

Table 1.4 gives mean values for six of the mandible measurements for specimens from all of the groups. The main question to be addressed here is how these groups are related and, in particular, how the prehistoric group is related to the others.

Example 1.5 Employment in European countries

Finally, as a contrast to the previous biological examples, consider the data in Table 1.5. This shows the percentages of the labour force in nine different types of industry for 26 European countries. In this case multivariate analyses may be useful in isolating groups of countries with similar employment distributions and in generally aiding the comprehension of the relationships between the countries.

Table 1.5 Percentages of people employed in nine different industry groups in Europe (AGR = agriculture, MIN = mining, MAN = manufacturing, PS = power supplies, CON = construction, SER = service industries, FIN = finance, SPS = social and personal services, TC = transport and communications).

Country	AGR	MIN	MAN	PS	CON	SER	FIN	SPS	TC
Belgium	3.3	0.9	27.6	0.9	8.2	19.1	6.2	26.6	7.2
Denmark	9.2	0.1	21.8	0.6	8.3	14.6	6.5	32.2	7.1
France	10.8	0.8	27.5	0.9	8.9	16.8	6.0	22.6	5.7
W. Germany	6.7	1.3	35.8	0.9	7.3	14.4	5.0	22.3	6.1
Ireland	23.2	1.0	20.7	1.3	7.5	16.8	2.8	20.8	6.1
Italy	15.9	0.6	27.6	0.5	10.0	18.1	1.6	20.1	5.7
Luxemburg	7.7	3.1	30.8	0.8	9.2	18.5	4.6	19.2	6.2
Netherlands	6.3	0.1	22.5	1.0	9.9	18.0	6.8	28.5	6.8
UK	2.7	1.4	30.2	1.4	6.9	16.9	5.7	28.3	6.4
Austria	12.7	1.1	30.2	1.4	9.0	16.8	4.9	16.8	7.0
Finland	13.0	0.4	25.9	1.3	7.4	14.7	5.5	24.3	7.6
Greece	41.4	0.6	17.6	0.6	8.1	11.5	2.4	11.0	6.7
Norway	9.0	0.5	22.4	0.8	8.6	16.9	4.7	27.6	9.4
Portugal	27.8	0.3	24.5	0.6	8.4	13.3	2.7	16.7	5.7
Spain	22.9	0.8	28.5	0.7	11.5	9.7	8.5	11.8	5.5
Sweden	6.1	0.4	25.9	0.8	7.2	14.4	6.0	32.4	6.8
Switzerland	7.7	0.2	37.8	0.8	9.5	17.5	5.3	15.4	5.7
Turkey	66.8	0.7	7.9	0.1	2.8	5.2	1.1	11.9	3.2
Bulgaria	23.6	1.9	32.3	0.6	7.9	8.0	0.7	18.2	6.7
Czechoslovakia	16.5	2.9	35.5	1.2	8.7	9.2	0.9	17.9	7.0
E. Germany	4.2	2.9	41.2	1.3	7.6	11.2	1.2	22.1	8.4
Hungary	21.7	3.1	29.6	1.9	8.2	9.4	0.9	17.2	8.0
Poland	31.1	2.5	25.7	0.9	8.4	7.5	0.9	16.1	6.9
Romania	34.7	2.1	30.1	0.6	8.7	5.9	1.3	11.7	5.0
USSR	23.7	1.4	25.8	0.6	9.2	6.1	0.5	23.6	9.3
Yugoslavia	48.7	1.5	16.8	1.1	4.9	6.4	11.3	5.3	4.0

Source: Euromonitor (1979, pp. 76–7) with the percentage employed in finance in Spain reduced from 14.7 to the more reasonable figure of 8.5.

1.2 Preview of multivariate methods

The five examples just considered are typical of the raw material for multivariate statistical methods. The main thing to note at this point is that in all cases there are several variables of interest and these are clearly not independent of each other. However, it is useful also to give a brief preview of what is to come in the chapters that follow in relationship to these examples.

Principal component analysis is designed to reduce the number of variables that need to be considered to a small number of indices (called the principal components) that are linear combinations of the original variables. For example, much of the variation in the body measurements of sparrows shown in Table 1.1 will be related to the general size of the birds, and the total

$$I_1 = X_1 + X_2 + X_3 + X_4 + X_5$$

will measure this quite well. This accounts for one 'dimension' in the data. Another index is

$$I_2 = X_1 + X_2 + X_3 - X_4 - X_5,$$

which is a contrast between the first three measurements and the last two. This reflects another 'dimension' in the data. Principal component analysis provides an objective way of finding indices of this type so that the variation in the data can be accounted for as concisely as possible. It may well turn out that two or three principal components provide a good summary of all the original variables. Consideration of the values of the principal components instead of the values of the original variables may then make it much easier to understand what the data have to say. In short, principal component analysis is a means of simplifying data by reducing the number of variables.

Factor analysis also attempts to account for the variation in a number of original variables using a smaller number of index variables or factors. It is assumed that each original variable can be expressed as a linear combination of these factors, plus a residual term that reflects the extent to which the variable is independent of the other variables. For example, a two-factor model for the sparrow data assumes that

$$X_1 = a_{11}F_1 + a_{12}F_2 + e_1$$
$$X_2 = a_{21}F_1 + a_{22}F_2 + e_2$$
$$X_3 = a_{31}F_1 + a_{32}F_2 + e_3$$
$$X_4 = a_{41}F_1 + a_{42}F_2 + e_4$$

and

$$X_5 = a_{51}F_1 + a_{52}F_2 + e_5,$$

where the a_{ij} values are constants, F_1 and F_2 are the factors, and e_i represents the variation in X_i that is independent of the variation in the other X-variables. Here F_1 might be the factor of size. In that case the coefficients a_{11}, a_{21}, a_{31}, a_{41} and a_{51} would all be positive, reflecting the fact that some birds tend to be large and some birds tend to be small on all body measurements. The second factor F_2 might then measure an aspect of the shape of birds, with some positive coefficients and some negative coefficients. If this two-factor model fitted the data well then it would provide a relatively straightforward description of the relationship between the five body measurements being considered.

One type of factor analysis starts by taking a few principal components as the factors in the data being considered. These initial factors are then modified by a special transformation process called 'factor rotation' in order to make them easier to interpret. Other methods for finding initial factors are also used. A rotation to simpler factors is almost always done.

Discriminant function analysis is concerned with the problem of seeing whether it is possible to separate different groups on the basis of the available measurements. This could be used, for example, to see how well surviving and non-surviving sparrows can be separated using their body measurements (Example 1.1), or how skulls from different epochs can be separated, again using size measurements (Example 1.2). Like principal component analysis, discriminant function analysis is based on the idea of finding suitable linear combinations of the original variables.

Cluster analysis is concerned with the identification of groups of similar individuals. There is not much point in doing this type of analysis with data like that of Examples 1.1 and 1.2 since the groups (survivors, non-survivors; epochs) are already known. However, in Example 1.3 there might be some interest in grouping colonies on the basis of environmental variables or Pgi frequencies, while in Example 1.4 the main point of interest is in the similarity between prehistoric dogs and other animals. Similarly, in Example 1.5 different European countries can be grouped in terms of similarity between employment patterns.

With *canonical correlation* the variables (not the individuals) are divided into two groups and interest centres on the relationship between these. Thus in Example 1.3 the first four variables are related to the environment while the remaining six variables reflect the genetic distribution at the different colonies of *Euphydryas editha*. Finding what relationships, if any, exist between these two groups of variables is of considerable biological interest.

Finally, there is *multidimensional scaling*. The method begins with data on some measure of the distances apart of a number of individuals. From these distances a 'map' is constructed showing how the individuals are related. This is a useful facility since it is often possible to measure how far apart pairs of objects are without having any idea of how the objects are related in a geometric sense. Thus in Example 1.4 there are ways of measuring the 'distances' between modern dogs and golden jackals, modern dogs and Chinese wolves, etc. Considering each pair of animal groups gives 21 distances altogether. From these distances multidimensional scaling can be used to produce a 'map' of the relationships between the groups. With a one-dimensional 'map' the groups are placed along a straight line. With a two-dimensional 'map' they are represented by points on a plane. With a three-dimensional 'map' they are represented by points within a cube. Four- and higher-dimensional solutions are also possible although these have limited use because they cannot be visualized. The value of a one-, two- or three-dimensional map is clear for Example 1.4 since such a map would immediately show which groups prehistoric dogs are similar to and which groups they are different from. Hence multidimensional scaling may be a useful alternative to cluster analysis in this case. A 'map' of European countries by employment patterns might also be of interest in Example 1.5.

1.3 The multivariate normal distribution

The normal distribution for a single variable should be familiar to readers of this book. It has the well known 'bell-shaped' frequency curve. Many standard univariate statistical methods are based on the assumption that data are normally distributed.

Knowing the prominence of the normal distribution with univariate statistical methods, it will come as no surprise to discover that the multivariate normal distribution has a central position with multivariate statistical methods. Many of these methods require the assumption that the data being analysed have multivariate normal distributions.

The exact definition of a multivariate normal distribution is not too important. The approach of most people, for better or worse, seems to be to regard data as being normally distributed unless there is some reason to believe that this is not true. In particular, if all the individual variables being studied appear to be normally distributed, then it is assumed that the joint distribution is multivariate normal. This is, in fact, a minimum requirement since the definition of multivariate normality requires more than this.

Cases do arise where the assumption of multivariate normality is clearly invalid. For example, one or more of the variables being studied may have a highly skewed distribution with several outlying high (or low) values; there may be many repeated values; etc. This type of problem can sometimes be overcome by an appropriate transformation of the data, as discussed in elementary texts on statistics. If this does not work then a rather special form of analysis may be required.

One important aspect of a multivariate normal distribution is that it is specified completely by a mean vector and a covariance matrix. The definitions of a mean vector and a covariance matrix are given in Section 2.7.

1.4 Computer programs

Practical methods for carrying out the calculations for multivariate analyses were developed from the mid-1930s. However, the application of these methods for more than small numbers of variables had to wait until computing equipment was sufficiently well developed. It is only in the last 20 years or so that analyses have become reasonably easy to carry out. Nowadays there are many standard statistical packages available for calculations, for example BMDP, SAS and SPSS. It is the intention that this book should provide readers with enough information to use any package intelligently, without saying much about any particular one.

Most multivariate analyses are still done using the standard packages on medium or large computers. However, the increasing availability and power of microcomputers suggests that this will not be the case for much longer. Packages will be 'shrunk down' to fit into micros and, also, special-purpose programs will become increasingly available in languages like BASIC. Indeed, it is not difficult to write BASIC programs to do many of the standard multivariate analyses providing advantage is taken of the availability of published algorithms to do the complicated parts of the calculations. Some limited instructions in this direction are included in the chapters that follow.

References

Bumpus, H.C. (1898) The elimination of the unfit as illustrated by the introduced sparrow, *Passer domesticus. Biological Lectures, Marine Biology Laboratory, Woods Hole*, 11th Lecture, pp. 209–26.

Euromonitor (1979) *European Marketing Data and Statistics*. Euromonitor Publications, London.

Higham, C.F.W., Kijngam, A. and Manly, B.F.J. (1980) An analysis of prehistoric canid remains from Thailand. *Journal of Archaeological Science* **7**, 149–65.

McKechnie, S.W., Ehrlich, P.R. and White, R.R. (1975) Population genetics of *Euphydryas* butterflies. I. Genetic variation and the neurality hypothesis. *Genetics* **81**, 571–94.

Thomson, A. and Randall-Maciver, R. (1905) *Ancient Races of the Thebaid*. Oxford University Press.

CHAPTER TWO

Matrix algebra

2.1 The need for matrix algebra

As indicated in the Preface, the theory of multivariate statistical methods can only be explained reasonably well with the use of some matrix algebra. For this reason it is helpful, if not essential, to have a certain minimal knowledge of this area of mathematics. This is true even for those whose interest is solely in using the methods as a tool. At first sight, the notation of matrix algebra is certainly somewhat daunting. However, it is not difficult to understand the basic principles involved providing that some details are accepted on faith.

2.2 Matrices and vectors

A *matrix* of size $m \times n$ is an array of numbers with m rows and n columns, considered as a single entity, of the form

$$\mathbf{A} = \begin{bmatrix} a_{11} & a_{12} & \cdots & a_{1n} \\ a_{21} & a_{22} & \cdots & a_{2n} \\ \vdots & \vdots & & \\ a_{m1} & a_{m2} & \cdots & a_{mn} \end{bmatrix}$$

If $m = n$, then this is a *square* matrix. If a matrix only has one column, for instance,

$$\mathbf{c} = \begin{bmatrix} c_1 \\ c_2 \\ \vdots \\ c_m \end{bmatrix},$$

then this is called a *column vector*. If there is only one row, for instance

$$\mathbf{r} = (r_1, r_2, \ldots, r_n),$$

then this is called a *row vector*.

The *transpose* of a matrix is found by interchanging the rows and columns. Thus the transpose of **A** above is

$$\mathbf{A}' = \begin{bmatrix} a_{11} & a_{21} & \cdots & a_{m1} \\ a_{12} & a_{22} & \cdots & a_{m2} \\ \vdots & \vdots & & \\ a_{1n} & a_{2n} & \cdots & a_{mn} \end{bmatrix}.$$

Also, $\mathbf{c}' = (c_1, c_2, \ldots, c_m)$, and $\mathbf{r}'$ is a column vector.

There are a number of special kinds of matrix that are particularly important. A *zero matrix* has all elements equal to zero:

$$\mathbf{0} = \begin{bmatrix} 0 & 0 & \cdots & 0 \\ 0 & 0 & \cdots & 0 \\ \vdots & & & \\ 0 & 0 & \cdots & 0 \end{bmatrix}.$$

A *diagonal matrix* is a square matrix that has elements of zero, except down the main diagonal, and so is of the form

$$\mathbf{T} = \begin{bmatrix} t_1 & 0 & 0 & \cdots & 0 \\ 0 & t_2 & 0 & \cdots & 0 \\ 0 & 0 & t_3 & \cdots & 0 \\ \vdots & & & & \\ 0 & 0 & 0 & \cdots & t_n \end{bmatrix}.$$

A *symmetric matrix* is a square matrix that is unchanged when it is transposed, so that **A** has this property providing that $\mathbf{A}' = \mathbf{A}$. Finally, an *identity matrix* is a diagonal matrix with all diagonal terms being unity:

$$\mathbf{T} = \begin{bmatrix} 1 & 0 & 0 & \cdots & 0 \\ 0 & 1 & 0 & \cdots & 0 \\ 0 & 0 & 1 & \cdots & 0 \\ \vdots & \vdots & \vdots & & \vdots \\ 0 & 0 & 0 & \cdots & 1 \end{bmatrix}.$$

Two matrices are *equal* only if all their elements agree. For example,

$$\begin{bmatrix} a_{11} & a_{12} & a_{13} \\ a_{21} & a_{22} & a_{23} \\ a_{31} & a_{32} & a_{33} \end{bmatrix} = \begin{bmatrix} b_{11} & b_{12} & b_{13} \\ b_{21} & b_{22} & b_{23} \\ b_{31} & b_{32} & b_{33} \end{bmatrix}$$

only if $a_{11} = b_{11}$, $a_{12} = b_{12}, \dots, a_{33} = b_{33}$.

The *trace* of a matrix is the sum of the diagonal terms. Thus $\text{tr}(\mathbf{A}) = a_{11} + a_{22} + \cdots + a_{nn}$ for an $n \times n$ matrix. This is only defined for square matrices.

2.3 Operations on matrices

The ordinary arithmetic processes of addition, subtraction, multiplication and division have their counterparts with matrices. With addition and subtraction it is simply a matter of working element by element. For example, if $\mathbf{A}$ and $\mathbf{D}$ are both 3×2 matrices, then

$$\mathbf{A} + \mathbf{D} = \begin{bmatrix} a_{11} & a_{12} \\ a_{21} & a_{22} \\ a_{31} & a_{32} \end{bmatrix} + \begin{bmatrix} d_{11} & d_{12} \\ d_{21} & d_{22} \\ d_{31} & d_{32} \end{bmatrix} = \begin{bmatrix} a_{11} + d_{11} & a_{12} + d_{12} \\ a_{21} + d_{21} & a_{22} + d_{22} \\ a_{31} + d_{31} & a_{32} + d_{32} \end{bmatrix}$$

while

$$\mathbf{A} - \mathbf{D} = \begin{bmatrix} a_{11} - d_{11} & a_{12} - d_{12} \\ a_{21} - d_{21} & a_{22} - d_{22} \\ a_{31} - d_{31} & a_{32} - d_{32} \end{bmatrix}.$$

These operations can only be carried out with two matrices that are of the same size.

In matrix algebra, an ordinary number such as 20 is called a *scalar*. Multiplication of a matrix $\mathbf{A}$ by a scalar k is then defined as multiplying every element of $\mathbf{A}$ by k. Thus if $\mathbf{A}$ is 3×2, as shown above, then

$$k\mathbf{A} = \begin{bmatrix} ka_{11} & ka_{12} \\ ka_{21} & ka_{22} \\ ka_{31} & ka_{32} \end{bmatrix}.$$

Multiplying two matrices together is more complicated. To begin with, $\mathbf{A} \cdot \mathbf{B}$ is only defined if the number of columns of $\mathbf{A}$ is equal to the number of rows of **B**. If this is the case, say with **A** of size $m \times n$ and **B** of size $n \times p$, then

$$\mathbf{A} \cdot \mathbf{B} = \begin{bmatrix} a_{11} & a_{12} & \cdots & a_{1n} \\ a_{21} & a_{22} & \cdots & a_{2n} \\ \vdots & & & \\ a_{m1} & a_{m2} & \cdots & a_{mn} \end{bmatrix} \times \begin{bmatrix} b_{11} & b_{12} & \cdots & b_{1p} \\ b_{21} & b_{22} & \cdots & b_{2p} \\ & & & \\ b_{n1} & b_{n2} & \cdots & b_{np} \end{bmatrix}$$

$$= \begin{bmatrix} \sum a_{1j} b_{j1} & \sum a_{1j} b_{j2} & \cdots & \sum a_{1j} b_{jp} \\ \sum a_{2j} b_{j1} & \sum a_{2j} b_{j2} & \cdots & \sum a_{2j} b_{jp} \\ \vdots & & & \\ \sum a_{mj} b_{j1} & \sum a_{mj} b_{j2} & \cdots & \sum a_{mj} b_{jp} \end{bmatrix},$$

where $\sum$ indicates the summation for all $j = 1, 2, \ldots, n$. Thus the element in the ith row and kth column of $\mathbf{A} \cdot \mathbf{B}$ is

$$\sum a_{ij} b_{jk} = a_{i1} b_{1k} + a_{i2k} b_{2k} + \cdots + a_{in} b_{nk}.$$

It is only when **A** and **B** are square that $\mathbf{A} \cdot \mathbf{B}$ and $\mathbf{B} \cdot \mathbf{A}$ are both defined. However, even if this is true, $\mathbf{A} \cdot \mathbf{B}$ and $\mathbf{B} \cdot \mathbf{A}$ are not generally equal. For example,

$$\begin{bmatrix} 2 & -1 \\ 1 & 1 \end{bmatrix} \cdot \begin{bmatrix} 1 & 1 \\ 0 & 1 \end{bmatrix} = \begin{bmatrix} 2 & 1 \\ 1 & 2 \end{bmatrix}$$

while

$$\begin{bmatrix} 1 & 1 \\ 0 & 1 \end{bmatrix} \cdot \begin{bmatrix} 2 & -1 \\ 1 & 1 \end{bmatrix} = \begin{bmatrix} 3 & 0 \\ 1 & 1 \end{bmatrix}.$$

2.4 Matrix inversion

Matrix inversion is analogous to the ordinary arithmetic process of division. Thus for a scalar k it is, of course, true that $k \times k^{-1} = 1$. In a similar way, if **A** is a square matrix then its *inverse* is $\mathbf{A}^{-1}$, where $\mathbf{A} \times \mathbf{A}^{-1} = \mathbf{I}$, this being the identity matrix. Inverse matrices are only defined for square matrices. However, all square matrices do not have an inverse. When $\mathbf{A}^{-1}$ exists, it is both a right and left inverse so that $\mathbf{A}^{-1} \mathbf{A} = \mathbf{A} \mathbf{A}^{-1} = \mathbf{I}$.

An example of an inverse matrix is

$$\begin{bmatrix} 2 & 1 \\ 1 & 2 \end{bmatrix}^{-1} = \begin{bmatrix} 2/3 & -1/3 \\ -1/3 & 2/3 \end{bmatrix}.$$

This can be verified by checking that

$$\begin{bmatrix} 2 & 1 \\ 1 & 2 \end{bmatrix} \times \begin{bmatrix} 2/3 & -1/3 \\ -1/3 & 2/3 \end{bmatrix} = \begin{bmatrix} 1 & 0 \\ 0 & 1 \end{bmatrix}.$$

Actually, the inverse of a 2×2 matrix, if it exists, can be determined easily. It is given by

$$\begin{bmatrix} a_{11} & a_{12} \\ a_{21} & a_{22} \end{bmatrix}^{-1} = \begin{bmatrix} a_{22}/\Delta & -a_{12}/\Delta \\ -a_{21}/\Delta & a_{11}/\Delta \end{bmatrix},$$

where $\Delta = a_{11}a_{22} - a_{12}a_{21}$. Here the scalar quantity Δ is called the *determinant* of the matrix. Clearly the inverse is not defined if $\Delta = 0$, since finding the elements of the inverse then requires a division by zero. For 3×3 and larger matrices the calculation of an inverse is a tedious business best done using a computer program.

Any square matrix has a determinant value that can be calculated by a generalization of the equation just given for the 2×2 case. If the determinant of a matrix is zero then the inverse does not exist, and vice versa. A matrix with a zero determinant is described as being *singular*.

Matrices sometimes arise for which the inverse is equal to the transpose. They are then described as being *orthogonal*. Thus **A** is orthogonal if $\mathbf{A}^{-1} = \mathbf{A}'$.

2.5 Quadratic forms

Let **A** be an $n \times n$ matrix and **x** be a column vector of length n. Then the expression

$$Q = \mathbf{x}'\mathbf{A}\mathbf{x}$$

is called a *quadratic form*. It is a scalar and can be expressed by the alternative equation

$$Q = \sum_{i=1}^{n} \sum_{j=1}^{n} x_i a_{ij} x_j,$$

where x_i is the element in the ith row of **x** and a_{ij} is the element in the ith row and the jth column of **A**.

2.6 Eigenvalues and vectors

Consider the set of equations

$$a_{11}x_1 + a_{12}x_2 + \cdots + a_{1n}x_n = \lambda x_1$$
$$a_{21}x_1 + a_{22}x_2 + \cdots + a_{2n}x_n = \lambda x_2$$
$$\vdots$$
$$a_{n1}x_1 + a_{n2}x_2 + \cdots + a_{nn}x_n = \lambda x_n$$

which can be written in matrix form as
$$\mathbf{Ax} = \lambda\mathbf{x},$$
or
$$(\mathbf{A} - \lambda\mathbf{I})\mathbf{x} = \mathbf{0},$$

where $\mathbf{I}$ is an $n \times n$ identity matrix and $\mathbf{0}$ is an $n \times 1$ zero vector. It can be shown that these equations can only hold for certain particular values of the scalar λ that are called the *latent roots* or *eigenvalues* of the matrix $\mathbf{A}$. There can be up to n of these roots. Given the ith latent root λ_i, the equations can be solved by arbitrarily setting $x_1 = 1$. The resulting vector of x values

$$\mathbf{x}_i = \begin{bmatrix} 1 \\ x_{2i} \\ x_{3i} \\ \vdots \\ x_{ni} \end{bmatrix}$$

(or any multiple of it) is called the ith *latent vector* or the ith *eigenvector* of the matrix $\mathbf{A}$. The sum of the eigenvalues of $\mathbf{A}$ is equal to the trace of $\mathbf{A}$.

Finding eigenvalues and eigenvectors is not a simple matter. Like finding a matrix inverse, it is a job best done on a computer.

2.7 Vectors of means and covariance matrices

Populations and samples for a single random variable are often summarized by mean values and variances. Thus suppose that a random sample of size n yields the sample values $x_1, x_2, \ldots, x_n$. Then the *sample mean* is

$$\bar{x} = \sum_{i=1}^{n} x_i/n$$

while the *sample estimate of variance* is

$$s^2 = \sum_{i=1}^{n} (x_i - \bar{x})^2/(n-1).$$

These are estimates of the corresponding population parameters – the *population mean* μ, and the *population variance* σ^2.

In a similar way, multivariate populations and samples can be summarized by mean vectors and covariance matrices. These are defined as follows. Suppose that there are p variables $X_1, X_2, \ldots, X_p$ and the values of these for the ith individual in a sample are $x_{i1}, x_{i2}, \ldots, x_{ip}$, respectively. Then the sample mean of variable j is

$$\bar{x}_j = \sum_{i=1}^{n} x_{ij}/n, \tag{2.1}$$

while the sample variance is

$$s_j^2 = \sum_{i=1}^{n} (x_{ij} - \bar{x}_j)^2/(n-1). \tag{2.2}$$

In addition, the sample *covariance* between variables j and k is defined as

$$c_{jk} = \sum_{i=1}^{n} (x_{ij} - \bar{x}_j)(x_{ik} - \bar{x}_k)/(n-1), \tag{2.3}$$

this being a measure of the extent to which the two variables are linearly related. The ordinary correlation coefficient for variables j and k, r_{jk}, say, is related to the covariance by the expression

$$r_{jk} = c_{jk}/(s_j s_k). \tag{2.4}$$

It is clear from equations (2.3) and (2.4) that $c_{kj} = c_{jk}, r_{kj} = r_{jk}$ and $r_{kk} = 1$.

The *vector of sample means* is calculated using equation (2.1):

$$\bar{\mathbf{x}} = \begin{bmatrix} \bar{x}_1 \\ \bar{x}_2 \\ \vdots \\ \bar{x}_p \end{bmatrix}. \tag{2.5}$$

This can be thought of as the centre of the sample. It is an estimate of the *population vector of means*

$$\boldsymbol{\mu} = \begin{bmatrix} \mu_1 \\ \mu_2 \\ \vdots \\ \mu_p \end{bmatrix}. \tag{2.6}$$

The matrix of variances and covariances

$$\mathbf{C} = \begin{bmatrix} c_{11} & c_{12} & \cdots & c_{1p} \\ c_{21} & c_{22} & \cdots & c_{2p} \\ \vdots & & & \\ c_{p1} & c_{p2} & \cdots & c_{pp} \end{bmatrix}, \tag{2.7}$$

where $c_{ii} = s_i^2$, is called the *sample covariance matrix*, or sometimes the *sample dispersion matrix*. This reflects the amount of variation in the sample and also the extent to which the p variables are correlated. It is an estimate of the population covariance matrix.

The matrix of correlations as defined by equation (2.4), is

$$\mathbf{R} = \begin{bmatrix} r_{11} & r_{12} & \cdots & r_{1p} \\ r_{21} & r_{22} & \cdots & r_{2p} \\ \vdots & \vdots & & \\ r_{p1} & r_{p2} & \cdots & r_{pp} \end{bmatrix} = \begin{bmatrix} 1 & r_{12} & \cdots & r_{1p} \\ r_{21} & 1 & \cdots & r_{2p} \\ \vdots & \vdots & & \\ r_{p1} & r_{p2} & \cdots & 1 \end{bmatrix}. \tag{2.8}$$

This is called the *sample correlation matrix*. Like **C**, this must be symmetric.

2.8 Further reading

A book by Causton (1977) gives a somewhat fuller introduction to matrix algebra than the one given here, but excludes latent roots and vectors. In addition to the chapter on matrix algebra, the other parts of Causton's book provide a good review of general mathematics. A more detailed account of matrix theory, still at an introductory level, is provided by Namboodiri (1984).

Those interested in learning more about matrix inversion and finding eigenvectors and eigenvalues, particularly methods for use on microcomputers, will find the book by Nash (1979) a useful source of information.

References

Causton, D.R. (1977) *A Biologist's Mathematics*. Edward Arnold, London.
Namboodiri, K. (1984) *Matrix Algebra: An Introduction*. Sage University Paper Series on Quantitative Applications in the Social Sciences, 07-038. Sage Publications, Beverly Hills.
Nash, J.C. (1979) *Compact Numerical Methods for Computers*. Adam Hilger, Bristol.

CHAPTER THREE

Tests of significance
with multivariate data

3.1 Introduction

The purpose of this chapter is to describe some tests that are available
for seeing whether there is any evidence that two or more samples
come from populations with different means or different amounts of
variation. To begin with, two-sample situations will be covered.

3.2 Comparison of mean values for two samples:
single variable case

Consider the data in Table 1.1 on the body measurements of 49
female sparrows. Consider in particular the first measurement, which
is total length. A question of some interest might be whether the mean
of this variable was the same for survivors and non-survivors of the
storm that led to the birds being collected. There is then a sample
(hopefully random) of 21 survivors and a second sample (again
hopefully random) of 28 non-survivors. We wish to know whether the
two sample means are significantly different. A standard approach
would be to carry out a t-test.

Thus, suppose that in a general situation there is a single variable X
and two random samples of values are available from different
populations. Let x_{i1} denote the values of X in the first sample, for
$i = 1, 2, \ldots, n_1$, and x_{i2} denote the values in the second sample, for
$i = 1, 2, \ldots, n_2$. Then the mean and variance for the jth sample are

$$\bar{x}_j = \sum_{i=1}^{n_j} x_{ij}/n_j$$

and

$$s_j^2 = \sum_{i=1}^{n} (x_{ij} - \bar{x}_j)^2/(n_j - 1). \qquad (3.1)$$

On the assumption that X is normally distributed in both samples, with a common within-sample variance, a test to see whether the two sample means are significantly different involves calculating the statistic

$$t = (\bar{x}_1 - \bar{x}_2) \bigg/ \left\{ s \sqrt{\left(\frac{1}{n_1} + \frac{1}{n_2}\right)} \right\} \qquad (3.2)$$

and seeing whether this is significantly different from zero in comparison with the t distribution with $n_1 + n_2 - 2$ degrees of freedom (d.f.). Here

$$s^2 = \{(n_1 - 1)s_1^2 + (n_2 - 1)s_2^2\}/(n_1 + n_2 - 2) \qquad (3.3)$$

is the pooled estimate of variance from the two samples.

It is known that this test is fairly robust to the assumption of normality. Providing that the population distributions of X are not too different from normal it should be satisfactory, particularly for sample sizes of about 20 or more. The assumption of equal within-sample variances is also not too crucial. Providing that the ratio of the true variances is within the limits 0.4 to 2.5, inequality of variance will have little adverse effect on the test. The test is particularly robust if the two sample sizes are equal, or nearly so (Carter *et al.*, 1979). If the population variances are very different then the t test can be modified to allow for this (Dixon and Massey, 1969, p. 119).

3.3 Comparison of mean values for two samples: multivariate case

Consider again the sparrow data of Table 1.1. The test described in the previous section can obviously be employed for each of the five measurements shown in the table (total length, alar extent, length of beak and head, length of humerus, and length of keel of sternum). In that way it is possible to decide which, if any, of these variables appear to have had different mean values for survivors and non-survivors. However, in addition to these it may also be of some interest to know whether all five variables considered together suggest a difference between survivors and non-survivors. In other words: does the total evidence point to mean differences between survivors and non-survivors?

What is needed to answer this question is a multivariate test. One possibility is Hotelling's T^2 test. The statistic used is then a generalization of the t statistic of equation (3.2) or, to be more precise, the square of the t statistic.

In a general case there will be p variables $X_1, X_2, \ldots, X_p$ being considered, and two samples with sizes n_1 and n_2. There are then two-sample mean vectors $\mathbf{x}_1$ and $\mathbf{x}_2$, with each one being calculated as shown in equations (2.1) and (2.5). There are also two-sample covariance matrices, $\mathbf{C}_1$ and $\mathbf{C}_2$, with each one being calculated as shown in equations (2.2), (2.3) and (2.7).

Assuming that the population covariance matrices are the same for both populations, a pooled estimate of this matrix is

$$\mathbf{C} = \{(n_1 - 1)\mathbf{C}_1 + (n_2 - 1)\mathbf{C}_2\}/(n_1 + n_2 - 2). \qquad (3.4)$$

Hotelling's T^2 statistic is defined as

$$T^2 = n_1 n_2 (\bar{\mathbf{x}}_1 - \bar{\mathbf{x}}_2)' \mathbf{C}^{-1} (\bar{\mathbf{x}}_1 - \bar{\mathbf{x}}_2)/(n_1 + n_2). \qquad (3.5)$$

A significantly large value for this statistic is evidence that the mean vectors are different for the two sampled populations. The significance or lack of significance of T^2 is most simply determined by using the fact that in the null hypothesis case of equal population means the transformed statistic

$$F = (n_1 + n_2 - p - 1)T^2 / \{(n_1 + n_2 - 2)p\} \qquad (3.6)$$

follows an F distribution with p and $(n_1 + n_2 - p - 1)$ d.f.

Since T^2 is a quadratic form it is a scalar which can be written in the alternative way

$$T^2 = \frac{n_1 n_2}{n_1 + n_2} \sum_{i=1}^{p} \sum_{k=1}^{p} (\bar{x}_{1i} - \bar{x}_{2i}) c^{ik} (\bar{x}_{1k} - \bar{x}_{2k}), \qquad (3.7)$$

which may be simpler to compute. Here $\bar{x}_{jl}$ is the mean of variable X_l in the jth sample and c^{ik} is the element in the ith row and kth column of the inverse matrix $\mathbf{C}^{-1}$.

Hotelling's T^2 statistic is based on an assumption of normality and equal within-sample variability. To be precise, the two samples being compared using the T^2 statistic are assumed to come from multivariate normal distributions with equal covariance matrices. Some deviation from multivariate normality is probably not serious. A moderate difference between population covariance matrices is also not too important, particularly with equal or nearly equal sample sizes (Carter *et al.*, 1979). If the two population covariance matrices are very different, and sample sizes are very different as well, then a modified test can be used (Yao, 1965).

Example 3.1 Testing mean values for Bumpus's female sparrows

As an example of the use of the univariate and multivariate tests that have been described for two samples, consider the sparrow data shown in Table 1.1. Here it is a question of whether there is any difference between survivors and non-survivors with respect to the mean values of five morphological characters.

First of all, tests on the individual variables can be considered, starting with X_1, the total length. The mean of this variable for the 21 survivors is $\bar{x}_1 = 157.38$ while the mean for the 28 non-survivors is $\bar{x}_2 = 158.43$. The corresponding sample variances are $s_1^2 = 11.05$ and $s_2^2 = 15.07$. The pooled variance from equation (3.3) is therefore

$$s^2 = (20 \times 11.05 + 27 \times 15.07)/47 = 13.36,$$

and the t statistic of equation (3.2) is

$$t = (157.38 - 158.43) \Big/ \sqrt{\left\{ 13.36 \left(\frac{1}{21} + \frac{1}{28} \right) \right\}} = -0.99,$$

with $n_1 + n_2 - 2 = 47$ d.f. This is not significantly different from zero at the 5% level so there is no evidence of a mean difference between survivors and non-survivors with regard to total length.

Table 3.1 summarizes the results of tests on all five of the variables in Table 1.1 taken individually. In no case is there any evidence of a mean difference between survivors and non-survivors.

For tests on all five variables considered together it is necessary to know the sample mean vectors and covariances matrices. The means are given in Table 3.1. The covariances matrices are defined by

Table 3.1 Comparison of mean values for survivors and non-survivors for Bumpus's female sparrows with variables taken one at a time.

Variable	Survivors $\bar{x}_1$	s_1^2	Non-survivors $\bar{x}_2$	s_2^2	t (47 d.f.)
Total length	157.38	11.05	158.43	15.07	−0.99
Alar extent	241.00	17.50	241.57	32.55	−0.39
Length beak & head	31.43	0.53	31.48	0.73	−0.20
Length humerus	18.50	0.18	18.45	0.43	0.33
Length keel of sternum	20.81	0.58	20.84	1.32	−0.10

equation (2.7). For the sample of 21 survivors,

$$\bar{\mathbf{x}}_1 = \begin{bmatrix} 157.381 \\ 241.000 \\ 31.433 \\ 18.500 \\ 20.810 \end{bmatrix} \text{ and } \mathbf{C}_1 = \begin{bmatrix} 11.048 & 9.100 & 1.557 & 0.870 & 1.286 \\ 9.100 & 17.500 & 1.910 & 1.310 & 0.880 \\ 1.557 & 1.910 & 0.531 & 0.189 & 0.240 \\ 0.870 & 1.310 & 0.189 & 0.176 & 0.133 \\ 1.286 & 0.880 & 0.240 & 0.133 & 0.575 \end{bmatrix}$$

For the sample of 28 non-survivors,

$$\bar{\mathbf{x}}_2 = \begin{bmatrix} 158.429 \\ 241.571 \\ 31.479 \\ 18.446 \\ 20.839 \end{bmatrix} \text{ and } \mathbf{C}_2 = \begin{bmatrix} 15.069 & 17.190 & 2.243 & 1.746 & 2.931 \\ 17.190 & 32.550 & 3.398 & 2.950 & 4.066 \\ 2.243 & 3.398 & 0.728 & 0.470 & 0.559 \\ 1.743 & 2.950 & 0.470 & 0.434 & 0.506 \\ 2.931 & 4.066 & 0.559 & 0.506 & 1.321 \end{bmatrix}.$$

The pooled sample covariance matrix is then

$$\mathbf{C} = (20\mathbf{C}_1 + 27\mathbf{C}_2)/47 = \begin{bmatrix} 13.358 & 13.748 & 1.951 & 1.373 & 2.231 \\ 13.748 & 26.146 & 2.765 & 2.252 & 2.710 \\ 1.951 & 2.765 & 0.645 & 0.350 & 0.423 \\ 1.373 & 2.252 & 0.350 & 0.324 & 0.347 \\ 2.231 & 2.710 & 0.423 & 0.347 & 1.004 \end{bmatrix}$$

where, for example, the element in the second row and third column is $(20 \times 1.910 + 27 \times 3.398)/47 = 2.765$.

The inverse of the matrix **C** is found to be

$$
\mathbf{C}^{-1} = \begin{bmatrix}
0.2061 & -0.0694 & -0.2395 & 0.0785 & -0.1969 \\
-0.0694 & 0.1234 & -0.0376 & -0.5517 & 0.0277 \\
-0.2395 & -0.0376 & 4.2219 & -3.2624 & -0.0181 \\
0.0785 & -0.5517 & -3.2624 & 11.4610 & -1.2720 \\
-0.1969 & 0.0277 & -0.0181 & -1.2720 & 1.8068
\end{bmatrix}
$$

This can be verified by evaluating the product $\mathbf{C} \times \mathbf{C}^{-1}$ and seeing that this is a unit matrix (apart from rounding errors).

Substituting the elements of $\mathbf{C}^{-1}$ and other values into equation (3.7) produces

$$
T^2 = \frac{21 \times 28}{21 + 28}[(157.381 - 158.429) \times 0.2061 \times (157.381 - 158.429)
$$
$$
- (157.318 - 158.429) \times 0.0694 \times (241.000 - 241.571)
$$
$$
+ \cdots + (20.810 - 20.839) \times 1.8068 \times (20.810 - 20.839)]
$$
$$
= 2.824.
$$

Using equation (3.6) this converts to an F statistic of

$$
F = (21 + 28 - 5 - 1) \times 2.824/\{(21 + 28 - 2) \times 5\} = 0.517,
$$

with 5 and 43 d.f. Clearly this is not significantly large since a significant F value must exceed unity. Hence there is no evidence of a difference in means for survivors and non-survivors, taking all five variables together.

3.4 Multivariate versus univariate tests

In this last example there were no significant results either for the variables considered individually or for the overall multivariate test. It should be noted, however, that it is quite possible to have insignificant univariate tests but a significant multivariate test. This can occur because of the accumulation of the evidence from the individual variables in the overall test. Conversely, an insignificant multivariate test can occur when some univariate tests are significant because the evidence of a difference provided by the significant variables is swamped by the evidence of no difference provided by the other variables.

One important aspect of the use of a multivariate test as distinct from a series of univariate tests concerns the control of type one error rates. A *type one error* involves finding a significant result when, in reality, the two samples being compared come from the same population. With a univariate test at the 5% level there is a 0.95 probability of a non-significant result when the population means are the same. Hence if p independent tests are carried out under these conditions then the probability of getting no significant results is 0.95^p. The probability of at least one significant result is therefore $1 - 0.95^p$. With many tests this can be quite a large probability. For example, if p is 5, the probability of at least one significant result by chance alone is $1 - 0.95^5 = 0.23$. With multivariate data, variables are usually not independent so $1 - 0.95^p$ does not quite give the correct probability of at least one significant result by chance alone if variables are tested one by one with univariate t tests. However, the principle still applies: the more tests that are made, the higher the probability of obtaining at least one significant result by chance.

On the other hand, a multivariate test such as Hotelling's T^2 test using the 5% level of significance gives a 0.05 probability of a type one error, irrespective of the number of variables involved. This is a distinct advantage over a series of univariate tests, particularly when the number of variables is large.

There are ways of adjusting significance levels in order to control the overall probability of a type one error when several tests are carried out. However, the use of a single multivariate test provides a better alternative procedure in many cases. A multivariate test has the added advantage of taking proper account of the correlation between variables.

3.5 Comparison of variation for two samples: single variable case

With a single variable, the best known method for comparing the variation in two samples is the F test. If s_j^2 is the variance in the jth sample, calculated as shown in equation (3.1), then the ratio s_1^2/s_2^2 is compared with percentage points of the F distribution with $(n_1 - 1)$ and $(n_2 - 1)$ d.f. Unfortunately, the F test is known to be rather sensitive to the assumption of normality. A significant result may well be due to the fact that a variable is not normally distributed rather than to unequal variances. For this reason it is sometimes argued that the *F* test should never be used to compare variances.

A robust alternative to the *F* test is Levene's (1960) test. The idea here is to transform the original data into absolute deviations from the mean and then test for a significant difference between the mean deviations in the two samples, using a *t* test. Absolute deviations from the arithmetic mean are usually used but a more robust test is possible by using absolute deviations from sample medians (Schultz, 1983). The procedure is illustrated in Example 3.2 below.

3.6 Comparison of variation for two samples: multivariate case

Most textbooks on multivariate methods suggest the use of Bartlett's test to compare the variation in two multivariate samples. This is described, for example, by Srivastava and Carter (1983, p. 333). However, this test is rather sensitive to the assumption that the samples are from multivariate normal distributions. There is always the possibility that a significant result is due to non-normality rather than to unequal population covariance matrices.

An alternative procedure that should be more robust can be constructed using the principle behind Levene's test. Thus the data values can be transformed into absolute deviations from sample means or medians. The question of whether two samples display significantly different amounts of variation is then transformed into a question of whether the transformed values show significantly different mean vectors. Testing of the mean vectors can be done using a T^2 test.

Another possibility was suggested by Van Valen (1978). This involves calculating

$$d_{ij} = \sqrt{\left\{ \sum_{k=1}^{p} (x_{ijk} - \bar{x}_{jk})^2 \right\}}, \tag{3.8}$$

where x_{ijk} is the value of variable X_k for the *i*th individual in sample *j*, and $\bar{x}_{jk}$ is the mean of the same variable in the sample. The sample means of the d_{ij} values are compared by a *t* test. Obviously if one sample is more variable than another then the mean d_{ij} value will be higher in the more variable sample.

To ensure that all variables are given equal weight, they should be standardized before the calculation of the d_{ij} values. Coding them to have unit variances will achieve this. For a more robust test it may be better to use sample medians in place of the sample means in equation (3.8). Then the formula for d_{ij} values is

$$d_{ij} = \sqrt{\left\{ \sum_{k=1}^{p} (x_{ijk} - M_{jk})^2 \right\}}, \qquad (3.9)$$

where M_{jk} is the median for variable X_k in the jth sample.

The T^2 test and Van Valen's test for deviations from medians are illustrated in the example that follows.

One point to note about the use of the test statistics (3.8) and (3.9) is that they are based on an implicit assumption that if the two samples being tested differ, then one sample will be more variable than the other for all variables. A significant result cannot be expected in a case where, for example, X_1 and X_2 are more variable in sample 1 but X_3 and X_4 are more variable in sample 2. The effect of the differing variances would then tend to cancel out in the calculation of d_{ij}. Thus Van Valen's test is not appropriate for situations where changes in the level of variation are not expected to be consistent for all variables.

Example 3.2 Testing variation for Bumpus's female sparrows

With Bumpus's data shown in Table 1.1, the most interesting question concerns whether the non-survivors were more variable than the survivors. This is what is expected if stabilizing selection took place.

First of all, the individual variables can be considered one at a time, starting with X_1, the total length. For Levene's test the original data values are transformed into deviations from sample medians. The median for survivors in 157 mm. The absolute deviations from this for the 21 birds in the sample then have a mean of $\bar{x}_1 = 2.57$ and a variance of $s_1^2 = 4.26$. The median for non-survivors is 159 mm. The absolute deviations from this median for the 28 birds in the sample have a mean of $\bar{x}_2 = 3.29$ with a variance of $s_2^2 = 4.21$. The pooled variance from equation (3.3) is 4.231 and the t statistic of equation (3.2) is

$$t = (2.57 - 3.29) \Big/ \sqrt{\left[4.231 \left\{ \frac{1}{21} + \frac{1}{28} \right\} \right]} = -1.21,$$

with 47 d.f.

Since non-survivors would be more variable than survivors if stabilizing selection occurred, it is a one-sided test that is required here, with low values of t providing evidence of selection. Clearly the observed value of t is not significantly low in the present instance. The t values for the other variables are as follows: alar extent, $t = -1.18$; length of beak and head, $t = -0.81$; length of humerus, $t = -1.91$; length of keel of sternum, $t = -1.40$. Only for the length of humerus is the result significantly low at the 5% level.

Table 3.2 shows absolute deviations from sample medians for the data after it has been standardized. For example, the first value given for variable 1, for survivors, is 0.28. This was obtained as follows. First, the original data were coded to have a zero mean and a unit variance for all 49 birds. This transformed the total length for the first survivor to $(156 - 157.98)/3.617 = -0.55$. The median transformed length for survivors was then -0.27. Hence the absolute deviation from the sample median for the first survivor is $|-0.55 - (-0.27)| = 0.28$, as recorded.

Comparing the transformed sample mean vectors for the five variables using Hotelling's T^2 test gives a test statistic of $T^2 = 4.75$, corresponding to an F statistic of 0.87 with 5 and 43 d.f. (equations (3.7) and (3.6)). There is therefore no evidence of a significant difference between the samples from this test.

Finally, consider Van Valen's test. The d values from equation (3.9) are shown in the last column of Table 3.2. The mean for survivors is 1.760, with variance 0.411. The mean for non-survivors is 2.265, with variance 1.133. The t value from equation (3.2) is then -1.92, which is significantly low at the 5% level. Hence this test indicates more variation for non-survivors than for survivors.

An explanation for the significant result with this test, but no significant result with the T^2 test, is not hard to find. As explained above, the T^2 test is not directional. Thus if the first sample has large means for some variables and small means for others when compared to the second sample, then all of the differences contribute to T^2. On the other hand, Van Valen's test is specifically for less variation in

Table 3.2 Absolute deviations from sample medians for Bumpus's female data and d values from equation (3.9).

Bird	Total length	Alar extent	Length beak & head	Length humerus	Length keel sternum	d
1	0.28	1.00	0.25	0.00	0.10	1.07
2	0.83	0.00	1.27	1.07	1.02	2.12
3	1.11	0.00	0.51	0.18	0.00	1.23
4	1.11	0.80	0.64	1.43	0.41	2.12
5	0.55	0.60	0.13	0.18	0.31	0.90
6	1.66	1.40	0.76	0.90	0.31	2.49
7	0.00	0.40	0.64	0.18	0.41	0.87
8	0.55	0.20	1.78	0.18	0.61	1.98
9	1.94	1.59	1.65	1.07	0.51	3.23
10	0.28	0.40	0.51	0.54	1.43	1.68
11	0.28	0.00	0.13	0.18	1.43	1.47
12	0.83	0.80	0.38	0.18	0.10	1.23
13	1.11	1.20	1.14	1.43	1.22	2.74
14	0.00	1.00	0.76	1.07	0.61	1.76
15	0.00	1.00	0.13	0.72	0.82	1.48
16	0.28	0.60	0.64	0.90	0.31	1.32
17	0.28	0.80	0.00	0.00	1.02	1.32
18	1.11	0.40	1.14	0.54	0.31	1.75
19	0.55	0.80	1.40	0.00	0.51	1.78
20	1.66	1.20	1.40	0.18	1.32	2.82
21	0.55	0.80	0.13	0.90	0.92	1.61
22	1.11	0.40	0.13	0.90	0.00	1.48
23	0.83	0.40	0.00	0.54	0.10	1.07
24	0.28	0.00	1.40	0.54	1.02	1.83
25	1.94	1.99	1.53	2.33	0.92	4.04
26	0.28	1.59	0.25	0.54	1.83	2.52
27	1.11	1.00	0.64	0.00	0.71	1.77
28	0.55	0.60	0.89	1.79	0.71	2.27
29	1.66	0.60	2.03	2.33	2.04	4.10
30	1.66	2.19	1.78	2.15	0.92	4.02
31	0.83	0.60	1.53	0.90	2.45	3.19
32	0.83	0.20	0.13	0.54	0.61	1.19
33	0.00	0.60	0.38	0.00	1.02	1.24
34	0.00	1.00	0.76	0.72	1.73	2.26
35	1.11	0.20	0.76	0.00	0.61	1.49
36	0.83	1.99	0.51	1.07	1.53	2.90
37	1.94	2.39	1.40	2.15	2.14	4.54
38	0.00	0.00	0.89	0.54	0.20	1.06
39	1.11	0.80	0.38	1.07	1.43	2.28
40	1.11	1.40	2.42	1.79	2.14	4.10
41	1.11	0.00	0.64	0.72	0.00	1.46

Table 3.2 (*Contd.*)

Bird	Total length	Alar extent	Length beak & head	Length humerus	Length keel sternum	d
42	0.83	1.00	0.25	0.54	0.41	1.48
43	0.00	0.80	0.00	0.18	0.41	0.91
44	0.55	0.60	0.76	1.07	0.10	1.55
45	1.11	1.40	1.02	1.43	1.12	2.74
46	0.83	1.00	0.51	1.07	0.31	1.79
47	1.66	1.00	1.14	0.18	0.31	2.28
48	0.83	0.60	1.27	0.00	0.41	1.68
49	1.38	1.20	1.02	0.54	0.20	2.17

sample 1 than in sample 2, for all variables. In the present case all of the variables show less variation in sample 1 than in sample 2. Van Valen's test has emphasized this fact but the T^2 test has not.

3.7 Comparison of means for several samples

When there is a single variable and several samples to be compared, the generalization of the *t* test is the *F* test from a one-factor analysis of variance. The calculations are as shown in Table 3.3.

When there are several variables and several samples, a so-called 'likelihood ratio test' can be used to compare the sample mean vectors. This involves calculating the statistic

$$\phi = [n - 1 - \tfrac{1}{2}(p - m)] \log_e[|\mathbf{T}|/|\mathbf{W}|] \qquad (3.10)$$

where *n* is the total number of observations, *p* is the number of variables, *m* is the number of samples, $|\mathbf{T}|$ is the determinant of the total sum of squares and cross-products matrix, and $|\mathbf{W}|$ is the determinant of the within-sample sum of squares and cross-products matrix. This statistic can be tested for significance by comparison with the chi-squared distribution with $p(m - 1)$ d.f. In Example 3.3 below the likelihood ratio test is used to compare the means for the five samples of male Egyptian skulls provided in Table 1.2.

The matrices **T** and **W** require some further explanation. Let x_{ijk} denote the value of variable X_k for the *i*th individual in the *j*th sample, $\bar{x}_{jk}$ denote the mean of X_k in the same sample, and $\bar{x}_k$ denote the

Table 3.3 One-factor analysis of variance for a single variable and m samples.

Source of variation	Sum of squares	Degrees of freedom	Mean square	F
Between samples	$B = T - W$	$m - 1$	$M_1 = B/(m-1)$	M_1/M_2
Within samples	$W = \sum_{j=1}^{m} \sum_{i=1}^{n_j} (x_{ij} - \bar{x}_j)^2$	$n - m$	$M_2 = W/(n-m)$	
Total	$T = \sum_{j=1}^{m} \sum_{i=1}^{n_j} (x_{ij} - \bar{x})^2$	$n - 1$		

$n_j =$ size of jth sample

$n = \sum_{j=1}^{m} n_j =$ total number of observations

$x_{ij} = i$th observation in jth sample

$\bar{x}_j = \sum_{i=1}^{n_j} x_{ij}/n_j =$ mean of jth sample

$\bar{x} = \sum_{j=1}^{m} \sum_{i=1}^{n_j} x_{ij}/n =$ overall mean

overall mean of X_k for all the data taken together. Then the element in row r and column c of $\mathbf{T}$ is

$$t_{rc} = \sum_{j=1}^{m} \sum_{i=1}^{n_j} (x_{ijr} - \bar{x}_r)(x_{ijc} - \bar{x}_c). \tag{3.11}$$

The element in the rth row and cth column of $\mathbf{W}$ is

$$w_{rc} = \sum_{j=1}^{m} \sum_{i=1}^{n_j} (x_{ijr} - \bar{x}_{jr})(x_{ijc} - \bar{x}_{jc}). \tag{3.12}$$

The test based on equation (3.10) involves the assumption that the distribution of the p variables is multivariate normal, with a constant within-sample covariance matrix. It is probably a fairly robust test in the sense that moderate deviations from this assumption do not unduly affect the characteristics of the test.

3.8 Comparison of variation for several samples

Bartlett's test is the best known for comparing the variation in several samples. This test has already been mentioned for the two-sample situation with several variables to be compared. See Srivastava and Carter (1983, p. 333) for details of the calculations involved. The test can be used with one or several variables. However, it does have the problem of being rather sensitive to deviations from normality in the distribution of the variables being considered.

Here, robust alternatives to Bartlett's test are recommended, these being generalizations of what was suggested for the two-sample situation. Thus absolute deviations from sample medians can be calculated for the data in m samples. For a single variable these can be treated as the observations for a one-factor analysis of variance. A significant F ratio is then evidence that the samples come from populations with different mean deviations, i.e., populations with different variability. If the ϕ statistic of equation (3.10) is calculated from the transformed data for p variables, then a significant result indicates that the covariance matrix is not constant for the m populations sampled.

Alternatively, the variables can be standardized to have unit variances for all the data lumped together and d values calculated using equation (3.9). These can then be analysed by a one-factor analysis of variance. This generalizes Van Valen's test that was suggested for comparing the variation in two multivariate samples. A significant F ratio from the analysis of variance indicates that some of the m populations sampled are more variable than others. As in the two-sample situation, this test is only really appropriate when some samples may be more variable than others for all the measurements being considered.

Example 3.3 Comparison of samples of Egyptian skulls

As an example of the test for comparing several samples, consider the data shown in Table 1.2 for four measurements on male Egyptian skulls for five samples of different ages.

A one-factor analysis of variance on the first variable, maximum breadth, provides $F = 5.95$, with 4 and 145 d.f. (Table 3.3). This is significantly large at the 0.1% level and hence there is clear evidence that the mean changed with time. For the other three variables, analysis of variance provides the following results: basibregmatic height, $F = 2.45$ (significant at the 5% level); basialveolar length $F = 8.31$ (significant at the 0.1% level); nasal height, $F = 1.51$ (not significant). It appears that the mean changed with time for the first three variables.

Next, consider the four variables together. If the five samples are combined then the matrix of sums of squares and products for the 150 observations, calculated using equation (3.11), is

$$\mathbf{T} = \begin{bmatrix} 3563.89 & -222.81 & -615.16 & 426.73 \\ -222.81 & 3635.17 & 1046.28 & 346.47 \\ -615.16 & 1046.28 & 4309.27 & -16.40 \\ 426.73 & 346.47 & -16.40 & 1533.33 \end{bmatrix},$$

for which the determinant is $|\mathbf{T}| = 7.306 \times 10^{13}$. The within-sample matrix of sums of squares and cross-products is found from equation (3.12) to be

$$\mathbf{W} = \begin{bmatrix} 3061.07 & 5.33 & 11.47 & 291.30 \\ 5.33 & 3405.27 & 754.00 & 412.53 \\ 11.47 & 754.00 & 3505.97 & 164.33 \\ 291.30 & 412.53 & 164.33 & 1472.13 \end{bmatrix},$$

for which the determinant is $|\mathbf{W}| = 4.848 \times 10^{13}$. Substituting $n = 150$,

$p = 4$, $m = 5$ and the values of $|\mathbf{T}|$ and $|\mathbf{W}|$ into equation (3.10) then yields $\phi = 61.31$, with $p(m - 1) = 16$ d.f. This is significantly large at the 0.1% level in comparison with the chi-square distribution. There is therefore clear evidence that the vector of mean values of the four variables changed with time.

For comparing the amount of variation in the samples it is a straightforward matter to transform the data into absolute deviations from sample medians. Analysis of variance then shows no significant difference between the sample means of the transformed data for any of the four variables. The ϕ statistic is not significant for all variables taken together. Also, analysis of variance shows no significant difference between the mean d values calculated using equation (3.9).

It appears that mean values changed with time for the four variables being considered but the variation about the means remained fairly constant.

3.9 Computational methods

The multivariate tests discussed in this chapter are not difficult to program on a microcomputer if standard algorithms are used where possible. The T^2 statistic of equation (3.5) requires a matrix inversion. This can be done easily using Algorithm 9 of Nash (1979). The ϕ test of equation (3.10) requires the calculation of two determinants. Algorithm 5 of Nash (1979) will provide these.

References

Carter, E.M., Khatri, C.G. and Srivastava, M.S. (1979) The effect of inequality of variances on the t-test. *Sankhya* **41**, 216–25.

Dixon, W.J. and Massey, F.J. (1969) *Introduction to Statistical Analysis* (3rd edn). McGraw-Hill, New York.

Levene, H. (1960) Robust tests for equality of variance. In *Contributions to Probability and Statistics* (Eds I. Olkin, S.G. Ghurye, W. Hoeffding, W.G. Madow and H.B. Mann), pp. 278–92. Stanford Univ. Press, California.

Nash, J.C. (1979) *Compact Numerical Methods for Computers*. Adam Hilger, Bristol.

Schultz, B. (1983) On Levene's test and other statistics of variation. *Evolutionary Theory* **6**, 197–203.

Srivastava, M.S. and Carter, E.M. (1983) *An Introduction to Applied Multivariate Statistics*. North-Holland, New York.

Van Valen, L. (1978) The statistics of variation. *Evolutionary Theory* **4**, 33–43. (Erratum *Evolutionary Theory* **4**, 202.)

Yao, Y. (1965) An approximate degrees of freedom solution to the multivariate Behrens–Fisher problem. *Biometrika* **52**, 139–47.

CHAPTER FOUR

Measuring and testing multivariate distances

4.1 Multivariate distances

A large number of multivariate problems can be viewed in terms of 'distances' between single observations, or between samples of observations, or between populations of observations. For example, considering the data in Table 1.4 on mandible measurements of dogs, wolves, jackals, cuons and dingos, it is sensible to ask how far one of these groups is from the other six groups. The idea then is that if two animals have similar mean mandible measurements then they are 'close', whereas if they have rather different mean measurements then they are 'distant' from each other. Throughout this chapter it is this concept of 'distance' that is being used.

A large number of distance measures have been proposed and used in multivariate analyses. Here only some of the most common ones will be mentioned. It is fair to say that measuring distances is a topic where a certain amount of arbitrariness seems unavoidable.

A possible situation is that there are n objects being considered, with a number of measurements being taken on each of these, and the measurements are of two types. For example, in Table 1.3 results are given for four environmental variables and six gene frequencies for 16 colonies of a butterfly. Two sets of distances can therefore be calculated between the colonies. One set can be environmental distances and the other set genetic distances. An interesting question is then whether there is a significant relationship between these two sets of distances. Mantel's (1967) test which is described in Section 4.5 is useful in this context.

4.2 Distances between individual observations

To begin the discussion on measuring distances, consider the simplest case where there are n individuals, each of which has values for p

variables $X_1, X_2, \ldots, X_p$. The values for individual i can then be denoted by $x_{i1}, x_{i2}, \ldots, x_{ip}$ and those for individual j by $x_{j1}, x_{j2}, \ldots, x_{jp}$. The problem is to measure the 'distance' between individual i and individual j.

If there are only $p = 2$ variables then the values can be plotted as shown in Fig. 4.1. Pythagoras' theorem then says that the length, d_{ij}, of the line joining the point for individual i to be the point for individual j (the *Euclidean distance*) is

$$d_{ij} = \sqrt{\{(x_{i1} - x_{j1})^2 + (x_{i2} - x_{j2})^2\}}.$$

With $p = 3$ variables the values can be taken as the coordinates in space for plotting the positions of individuals i and j (Fig. 4.2). Pythagoras' theorem then gives the distance between the two points

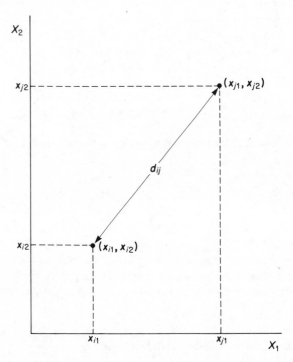

Figure 4.1 The Euclidean distance between individuals i and j, with $p = 2$ variables.

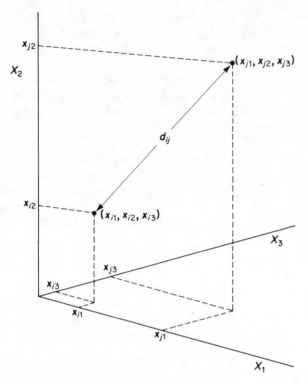

Figure 4.2 The Euclidean distance between individuals i and j, with $p = 3$ variables.

as

$$d_{ij} = \sqrt{\left\{ \sum_{k=1}^{3} (x_{ik} - x_{jk})^2 \right\}}.$$

With more than three variables it is not possible to use variable values as the coordinates for physically plotting points. However, the two- and three-variable cases suggest that the generalized Euclidean distance

$$d_{ij} = \sqrt{\left\{ \sum_{k=1}^{p} (x_{ik} - x_{jk})^2 \right\}} \tag{4.1}$$

may serve as a satisfactory measure for many purposes.

From the form of equation (4.1) it is clear that if one of the variables measured is much more variable than the others then this will dominate the calculation of distances. For example, to take an extreme case, suppose that n men are being compared and that X_1 is their stature and the other variables are tooth dimensions, with all the measurements being in millimetres. Stature differences will then be in the order of perhaps 20 or 30 millimetres while tooth dimension differences will be in the order of one or two millimetres. Simple calculations of d_{ij} will then provide distances between individuals that are essentially stature differences only, with tooth differences having negligible effects. Clearly there will be a scaling problem.

In practice it is usually desirable for all variables to have about the same influence on the distance calculation. This is achieved by a preliminary scaling of the variables to standardize them. This can be done, for example, by dividing each variable by its standard deviation for the n individuals being compared.

Example 4.1 Distances between dogs and related species

As an example of the use of the Euclidean distance measure, consider the data in Table 1.4 for mean mandible measurements of seven groups of dogs and related species. It may be recalled from Chapter 1 that the main question with these data is how the prehistoric dogs relate to the other groups.

The first step in calculating distances is to standardize the measurements. Here this will be done by expressing them as deviations from means in units of standard deviations. For example, the first measurement X_1 (breadth) has a mean of 10.486 mm and a

Table 4.1 Standardized variable values calculated from the original data in Table 1.4.

X_1	X_2	X_3	X_4	X_5	X_6
−0.50	−0.50	−0.74	−0.74	−0.49	−0.61
−1.52	−1.93	−1.12	−1.39	−0.86	−1.31
1.92	1.60	1.84	1.95	1.68	1.62
0.65	0.60	1.04	0.74	1.26	0.95
0.14	0.33	−0.04	0.00	−1.19	−0.40
−0.56	0.03	−0.14	−0.19	0.03	0.66
−0.12	−0.13	−0.84	−0.37	−0.43	−0.90

Table 4.2 Euclidean distances between seven animal groups.

	Modern dog	Golden jackal	Chinese wolf	Indian wolf	Cuon	Dingo	Prehistoric dog
Modern dog	—						
Golden jackal	2.07	—					
Chinese wolf	5.81	7.69	—				
Indian wolf	3.66	5.49	2.31	—			
Cuon	1.63	3.45	4.94	3.14	—		
Dingo	1.68	3.44	4.55	2.37	1.80	—	
Prehistoric dog	0.72	2.58	5.52	3.49	1.38	1.84	—

standard deviation of 1.572 mm for the seven groups. The standardized variable values are then calculated as follows: modern dog, $(9.7 - 10.486)/1.572 = -0.50$; golden jackal, $(8.1 - 10.486)/1.572 = -1.52;...$; prehistoric dog, $(10.3 - 10.486)/1.572 = -0.12$. Standardized values for all the variables are shown in Table 4.1.

Using equation (4.1) the distances shown in Table 4.2 have been calculated from the standardized variables. It is clear that the prehistoric dogs are rather similar to modern dogs in Thailand. Indeed, the distance between these two groups is the smallest distance in the whole table. (Higham *et al.* (1980) concluded from a more complicated analysis that the modern and prehistoric dogs are indistinguishable.)

4.3 Distances between populations and samples

A number of measures have been proposed for the distance between two multivariate populations when information is available on the means, variances and covariances of the populations. Here only two will be considered.

Suppose that g populations are available and the multivariate distributions in these populations are known for p variables $X_1, X_2, ..., X_p$. Let the mean of variable X_k in the ith population be μ_{ki}, and assume that the variance of X_k is the same value, V_k, in all the populations. Penrose (1953) proposed the relatively simple measure

$$P_{ij} = \sum_{k=1}^{p} \frac{(\mu_{ki} - \mu_{kj})^2}{pV_k} \tag{4.2}$$

for the distance between population i and population j.

A disadvantage of Penrose's measure is that it does not take into account the correlations between the p variables. This means that when two variables are measuring essentially the same thing, and hence are highly correlated, they still individually both contribute about the same amount to population distances as a third variable that is independent of all other variables.

A measure that does take into account correlations between variables is the Mahalanobis (1948) distance,

$$D_{ij}^2 = \sum_{r=1}^{p} \sum_{s=1}^{p} (\mu_{ri} - \mu_{rj}) v^{rs} (\mu_{si} - \mu_{sj}), \tag{4.3}$$

where v^{rs} is the element in the rth row and sth column of the inverse of the covariance matrix for the p variables. This is a quadratic form that can be written in the alternative way

$$D_{ij}^2 = (\boldsymbol{\mu}_i - \boldsymbol{\mu}_j)'\mathbf{V}^{-1}(\boldsymbol{\mu}_i - \boldsymbol{\mu}_j), \tag{4.4}$$

where

$$\boldsymbol{\mu}_i = \begin{bmatrix} \mu_{1i} \\ \mu_{2i} \\ \vdots \\ \mu_{pi} \end{bmatrix}$$

is the vector of means for the ith population and $\mathbf{V}$ is the covariance matrix. This measure can only be calculated if the population covariance matrix is the same for all populations.

The Mahalanobis distance is frequently used to measure the distance of a single multivariate observation from the centre of the population that the observation comes from. If $x_1, x_2, \ldots, x_p$ are the values of $X_1, X_2, \ldots, X_p$ for the individual, with corresponding population mean values of $\mu_1, \mu_2, \ldots, \mu_p$, then

$$D^2 = \sum_{r=1}^{p} \sum_{s=1}^{p} (x_r - \mu_r)v^{rs}(x_s - \mu_s)$$

$$= (\mathbf{x} - \boldsymbol{\mu})'\mathbf{V}^{-1}(\mathbf{x} - \boldsymbol{\mu}), \tag{4.5}$$

where $\mathbf{x}' = (x_1, x_2, \ldots, x_p)$ and $\boldsymbol{\mu}' = (\mu_1, \mu_2, \ldots, \mu_p)$. As before, $\mathbf{V}$ denotes the population covariance matrix and v^{rs} is the element in the rth row and sth column of $\mathbf{V}^{-1}$.

The value of D^2 can be thought of as a multivariate residual for the observation $\mathbf{x}$. A 'residual' here means a measure of how far the observation $\mathbf{x}$ is from the centre of the distributions of all values, taking into account all the variables being considered. An important and useful result is that if the population being considered is multivariate normally distributed, then the values of D^2 will follow a chi-square distribution with p degrees of freedom. A significantly large value of D^2 means that the corresponding observation is either (a) a genuine but unlikely record, or (b) a record containing some mistake. This suggests that a check should be made to see that the observation is correct and does not include an error.

The equations (4.2) to (4.5) can obviously be used with sample data if estimates of population means, variances and covariances are used in place of true values. In that case the covariance matrix $\mathbf{V}$ involved in equations (4.3) and (4.4) should be replaced with the pooled estimate from all the samples available. To be precise, suppose that there are m samples, with the ith sample being of size n_i, with a sample covariance matrix of $\mathbf{C}_i$. Then it is appropriate to take

$$\mathbf{C} = \sum_{i=1}^{m} (n_i - 1)\mathbf{C}_i \Big/ \sum_{i=1}^{m} (n_i - 1) \qquad (4.6)$$

as the pooled estimate of the common covariance matrix. The single-sample covariance matrix $\mathbf{C}_i$ is said to have $n_i - 1$ degrees of freedom, while $\mathbf{C}$ has a total of $\sum (n_i - 1)$ degrees of freedom. The sample covariance matrices should be calculated using equations (2.2) to (2.7).

In principle the Mahalanobis distance is superior to the Penrose distance because it uses information on covariances. However, this advantage is only present when covariances are accurately known. When covariances can only be estimated with a small number of degrees of freedom it is probably best to use the simpler Penrose measure. It is difficult to say precisely what a 'small number of degrees of freedom' means in this context. Certainly there should be no problem with using Mahalanobis distances based on a covariance matrix with the order of 100 or more degrees of freedom.

Example 4.2 Distances between samples of Egyptian skulls

For the five samples of male Egyptian skulls shown in Table 1.2 the mean vectors are as follows:

$$\bar{\mathbf{x}}_1 = \begin{bmatrix} 131.37 \\ 133.60 \\ 99.17 \\ 50.53 \end{bmatrix}, \bar{\mathbf{x}}_2 = \begin{bmatrix} 132.37 \\ 132.70 \\ 99.07 \\ 50.23 \end{bmatrix}, \bar{\mathbf{x}}_3 = \begin{bmatrix} 134.47 \\ 133.80 \\ 96.03 \\ 50.57 \end{bmatrix},$$

$$\bar{\mathbf{x}}_4 = \begin{bmatrix} 135.50 \\ 132.30 \\ 94.53 \\ 51.97 \end{bmatrix}, \text{ and } \bar{\mathbf{x}}_5 = \begin{bmatrix} 136.17 \\ 130.33 \\ 93.50 \\ 51.37 \end{bmatrix},$$

while the covariance matrices, calculated as indicated by equation (2.7), are

$$\mathbf{C}_1 = \begin{bmatrix} 26.31 & 4.45 & 0.45 & 7.25 \\ 4.15 & 19.97 & -0.79 & 0.39 \\ 0.45 & -0.79 & 34.63 & -1.92 \\ 7.25 & 0.39 & -1.92 & 7.64 \end{bmatrix},$$

$$\mathbf{C}_2 = \begin{bmatrix} 23.14 & 1.01 & 4.77 & 1.84 \\ 1.01 & 21.60 & 3.37 & 5.62 \\ 4.77 & 3.37 & 18.89 & 0.19 \\ 1.84 & 5.62 & 0.19 & 8.74 \end{bmatrix}$$

$$\mathbf{C}_3 = \begin{bmatrix} 12.12 & 0.79 & -0.78 & 0.90 \\ 0.79 & 24.79 & 3.59 & -0.09 \\ -0.78 & 3.59 & 20.72 & 1.67 \\ 0.90 & -0.09 & 1.67 & 12.60 \end{bmatrix},$$

$$\mathbf{C}_4 = \begin{bmatrix} 15.36 & -5.53 & -2.17 & 2.05 \\ -5.53 & 26.36 & 8.11 & 6.15 \\ -2.17 & 8.11 & 21.09 & 5.33 \\ 2.05 & 6.15 & 5.33 & 7.96 \end{bmatrix}$$

$$\text{and } \mathbf{C}_5 = \begin{bmatrix} 28.63 & -0.23 & -1.88 & -1.99 \\ -0.23 & 24.71 & 11.72 & 2.15 \\ -1.88 & 11.72 & 25.57 & 0.40 \\ -1.99 & 2.15 & 0.40 & 13.83 \end{bmatrix}.$$

Although the five sample covariance matrices appear to differ somewhat, it has been shown in Example 3.3 that the differences are not significant. It is therefore reasonable to pool them using equation (4.6). Since the sample sizes are all 30 this just amounts to taking the average of the five matrices, which is

$$\mathbf{C} = \begin{bmatrix} 21.111 & 0.037 & 0.079 & 2.009 \\ 0.037 & 23.485 & 5.200 & 2.845 \\ 0.079 & 5.200 & 24.179 & 1.133 \\ 2.009 & 2.845 & 1.133 & 10.153 \end{bmatrix}.$$

with $\sum(n_i - 1) = 145$ degrees of freedom.

Penrose's distance measures of equation (4.2) can now be calculated between each pair of samples. There are $p = 4$ variables with
variances that are estimated by $\hat{V}_1 = 21.111$, $\hat{V}_2 = 23.485$, $\hat{V}_3 = 24.179$
and $\hat{V}_4 = 10.153$, these being the diagonal terms in the pooled
covariance matrix. The sample mean values given in the vectors $\bar{\mathbf{x}}_1$ to
$\bar{\mathbf{x}}_5$ are estimates of population means. For example, the distance
between sample 1 and sample 2 is calculated as

$$P_{12} = \frac{(131.37 - 132.37)^2}{4 \times 21.111} + \frac{(133.60 - 132.70)^2}{4 \times 23.485}$$
$$+ \frac{(99.17 - 99.07)^2}{4 \times 24.179} + \frac{(50.53 - 50.23)^2}{4 \times 10.153}$$
$$= 0.23.$$

This only has meaning in comparison with the distances between the
other pairs of samples. Calculating these as well provides the
following distance matrix:

	Early pre-dynastic	Late pre-dynastic	12/13th dynasties	Ptolemaic	Roman
Early predynastic	—				
Late predynastic	0.023	—			
12/13th dynasties	0.216	0.163	—		
Ptolemaic	0.493	0.404	0.108	—	
Roman	0.736	0.583	0.244	0.066	—

It will be recalled from Example 3.3 that the mean values change
significantly from sample to sample. The Penrose distances show that
the changes are cumulative over time: the samples that are closest in
time are relatively similar whereas the samples that are far apart in
time are very different.

Turning next to Mahalanobis distances, these can be calculated
from equation (4.3). The inverse of the pooled covariance matrix $\mathbf{C}$ is

$$\mathbf{C}^{-1} = \begin{bmatrix} 0.0483 & 0.0011 & 0.0001 & -0.0099 \\ 0.0011 & 0.0461 & -0.0094 & -0.0121 \\ 0.0001 & -0.0094 & 0.0435 & -0.0022 \\ -0.0099 & -0.0121 & -0.0022 & 0.1041 \end{bmatrix}.$$

Using this and the sample means gives the distance from sample 1 to sample 2 to be

$$\begin{aligned}
D_{12}^2 = &(131.37 - 132.37)0.0483(131.37 - 132.37) \\
&+ (131.37 - 132.37)0.0011(133.60 - 132.70) \\
&+ \cdots - (50.53 - 50.23)0.0022(99.17 - 99.07) \\
&+ (50.53 - 50.23)0.1041(50.53 - 50.23) \\
= &\, 0.091.
\end{aligned}$$

Calculating the other distances between samples in the same way provides the distance matrix:

	Early pre- dynastic	Late pre- dynastic	12/13th dynasties	Ptolemaic	Roman
Early predynastic	—				
Late predynastic	0.091	—			
12/13th dynasties	0.903	0.729	—		
Ptolemaic	1.881	1.594	0.443	—	
Roman	2.697	2.176	0.911	0.219	—

A comparison between these distances and the Penrose distances shows a very good agreement. The Mahalanobis distances are three to four times as great as the Penrose distances. However, the relative distances between samples are almost the same for both measures. For example, the Penrose measure suggest that the distance from the early predynastic sample to the Roman sample is $0.736/0.023 = 32.0$ times as great as the distance from the early predynastic to the late predynastic sample. The corresponding ratio for the Mahalanobis measure is $2.697/0.091 = 29.6$.

4.4 Distances based upon proportions

A particular situation that sometimes occurs is that the variables being used to measure the distance between populations or samples are proportions whose sum is unity. For example, the animals of a certain species might be classified into K genetic classes. One colony might then have proportions p_1 of class 1, p_2 of class 2,...,p_K of class K, while a second colony has proportions q_1 of class 1, q_2 of class 2,...,q_k of class K. The question then arises of how similar the colonies are in genetic terms.

Various indices of distance have been proposed with this type of proportion data. For example,

$$d_1 = \sum_{i=1}^{K} |p_i - q_i|/2, \tag{4.7}$$

which is half of the sum of absolute proportion differences, is one possibility. This takes the value 1 when there is no overlap of classes and the value 0 when $p_i = q_i$ for all i. Another possibility is

$$d_2 = 1 - \sum_{i=1}^{K} p_1 q_i \bigg/ \sqrt{\left\{ \sum_{i=1}^{K} p_i^2 \sum_{i=1}^{K} q_i^2 \right\}}, \tag{4.8}$$

which again varies from 1 (no overlap) to 0 (equal proportions).

The merits of these and other distance measures for proportion data have been debated at length in the scientific literature. Here all that needs to be noted is that a large number of alternative measures exist. It must be hoped that for particular applications it does not matter much which one is used.

4.5 The Mantel test on distance matrices

A rather useful test for comparing two distance matrices was introduced by Mantel (1967) as a solution to the problem of detecting space and time clustering of diseases.

To understand the nature of the procedure the following simple example is helpful. Suppose that four objects are being studied, and that two sets of variables have been measured for each of these. The first set of variables can then be used to construct a 4×4 matrix where the entry in the ith row and jth column is the 'distance' between object i and object j. This distance matrix might be, for example,

$$\mathbf{M} = \begin{bmatrix} m_{11} & m_{12} & m_{13} & m_{14} \\ m_{21} & m_{22} & m_{23} & m_{24} \\ m_{31} & m_{32} & m_{33} & m_{34} \\ m_{41} & m_{42} & m_{43} & m_{44} \end{bmatrix} = \begin{bmatrix} 0.0 & 1.0 & 1.4 & 0.9 \\ 1.0 & 0.0 & 1.1 & 1.6 \\ 1.4 & 1.1 & 0.0 & 0.7 \\ 0.9 & 1.6 & 0.7 & 0.0 \end{bmatrix}$$

It is symmetric because, for example, the distance from object 2 to object 3 must be the same as the distance from object 3 to object 2

(1.1 units). Diagonal elements are zero since these represent distances from objects to themselves.

The second set of variables can also be used to construct a matrix of distances between the objects. For the example this will be taken as

$$\mathbf{E} = \begin{bmatrix} e_{11} & e_{12} & e_{13} & e_{14} \\ e_{21} & e_{22} & e_{23} & e_{24} \\ e_{31} & e_{32} & e_{33} & e_{34} \\ e_{41} & e_{42} & e_{43} & e_{44} \end{bmatrix} = \begin{bmatrix} 0.0 & 0.5 & 0.8 & 0.6 \\ 0.5 & 0.0 & 0.5 & 0.9 \\ 0.8 & 0.5 & 0.0 & 0.4 \\ 0.6 & 0.9 & 0.4 & 0.0 \end{bmatrix}$$

Like $\mathbf{M}$, this is symmetric with zeros down the diagonal.

Mantel's test is concerned with assessing whether the elements in $\mathbf{M}$ and $\mathbf{E}$ show correlation. Assuming $n \times n$ matrices, the test statistic

$$Z = \sum_{i=2}^{n} \sum_{j=1}^{i-1} m_{ij} e_{ij} \tag{4.9}$$

is calculated and compared with the distribution of Z that is obtained by taking the objects in a random order for one of the matrices. That is to say, matrix $\mathbf{M}$ can be left as it is. A random order can then be chosen for the objects for matrix $\mathbf{E}$. For example, suppose that a random ordering of objects turns out to be 3, 2, 4, 1. This then gives a randomized $\mathbf{E}$ matrix of

$$\mathbf{E}_{\mathbf{R}} = \begin{bmatrix} 0.0 & 0.5 & 0.4 & 0.8 \\ 0.5 & 0.0 & 0.9 & 0.5 \\ 0.4 & 0.9 & 0.0 & 0.6 \\ 0.8 & 0.5 & 0.6 & 0.0 \end{bmatrix}$$

The entry in row 1, column 2 is 0.5, the distance between objects 3 and 2; the entry in row 1, column 3 is 0.4, the distance between objects 3 and 4; and so on. A Z value can be calculated using $\mathbf{M}$ and $\mathbf{E}_{\mathbf{R}}$. Repeating this procedure using different random orders of the objects for $\mathbf{E}_{\mathbf{R}}$ produces the randomized distribution of Z. A check can then be made to see whether the observed Z value is a typical value from this distribution.

The basic idea is that if the two measures of distance are quite unrelated then the matrix $\mathbf{E}$ will be just like one of the randomly ordered matrices $\mathbf{E_R}$. Hence the observed Z will be a typical randomized Z value. On the other hand, if the two distance measures have a positive correlation then the observed Z will tend to be larger than values given by randomization. A negative correlation between distances should not occur but if it does then the result will be that the observed Z value will tend to be low when compared to the randomized distribution.

With n objects there are $n!$ different possible orderings of the object numbers. There are therefore $n!$ possible randomizations of the elements of $\mathbf{E}$, some of which might give the same Z values. Hence in our example with four objects the randomized Z distribution has $4! = 24$ equally likely values. It is not too difficult to calculate all of these. More realistic cases might involve, say, ten objects, in which case the number of possible Z values is $10! = 3,628,800$. Enumerating all of these then becomes impractical and there are two possible approaches for carrying out the Mantel test. A large number of randomized $\mathbf{E_R}$ matrices can be generated on the computer and the resulting distribution of Z values used in place of the true randomized distribution. Alternatively, the mean, $E(Z)$, and variance var(Z), of the randomized distribution of Z can be calculated, and

$$g = [Z - E(Z)]/[\text{var}(Z)]^{1/2}$$

can be treated as a standard normal variate.

Mantel (1967) provided formulae for the mean and variance of Z in the null hypothesis case of no correlation between the distance measures. There is, however, some doubt about the validity of the normal approximation for the test statistic g (Mielke, 1978). Given the ready availability of computers it therefore seems best to perform randomizations rather than to rely on this approximation.

The test statistic Z of equation (4.9) is the sum of the products of the elements in the lower diagonal parts of the matrices $\mathbf{M}$ and $\mathbf{E}$. The only reason for using this particular statistic is that Mantel's equations for the mean and variance are available. However, if it is decided to determine significance by computer randomizations there is no particular reason why the test statistic should not be changed. Indeed, values of Z are not particularly informative except in comparison with the mean and variance. It may therefore be more useful to take the correlation between the lower diagonal elements of

M and **E** as the test statistic instead of Z. This correlation is

$$r = \frac{Z - n(n-1)\bar{m}\bar{e}/2}{\sqrt{\left\{ \left(\sum_{i=2}^{n} \sum_{j=1}^{i-1} m_{ij}^2 - n(n-1)\bar{m}^2/2 \right) \left(\sum_{i=2}^{n} \sum_{j=1}^{i-1} e_{ij}^2 - n(n-1)\bar{e}^2/2 \right) \right\}}}$$

(4.10)

where $\bar{m}$ is the mean of the m_{ij} values, $\bar{e}$ is the mean of the e_{ij} values, and $n(n-1)/2$ is the number of lower diagonal elements in the matrices. This statistic has the usual interpretation in terms of the relationship between the two distance measures. Thus r lies in the range -1 to $+1$, with $r = -1$ indicating a perfect negative correlation, $r = 0$ indicating no correlation, and $r = +1$ indicating a perfect positive correlation. The significance or otherwise of the data will be the same for the test statistics Z and r since r is just Z with a constant subtracted, divided by another constant.

Example 4.3 More on distances between samples of Egyptian skulls

Returning to the Egyptian skull data, we can ask the question of whether the distances given in Example 4.2, based upon four skull measurements, are significantly related to the time differences between the five samples. This certainly does seem to be the case but a definitive answer is provided by Mantel's test.

The sample times are approximately 4000 BC (early predynastic), 3300 BC (late predynastic), 1850 BC (12th and 13th dynasties), 200 BC (Ptolemaic), and AD 150 (Roman). Comparing Penrose's distance measures with time differences (in thousands of years) therefore provides the following lower diagonal distance matrices between the samples:

Penrose's distances					*Times distances*				
—					—				
0.023	—				0.70	—			
0.216	0.163	—			2.15	1.45	—		
0.493	0.404	0.108	—		3.80	3.10	1.65	—	
0.736	0.583	0.244	0.066	—	4.15	3.45	2.00	0.35	—

The correlation between the elements of these matrices is 0.954. It appears, therefore, that the distances agree very well.

When 1000 random orderings of the time difference matrix were made by computer, only 19 of them resulted in a correlation of 0.954 or more. Consequently, the observed correlation can be declared significantly large at about the 2% (19/1000) level. There is clear evidence of a relationship between the two distance measures. A one-sided test is appropriate since there is no reason why skulls should become more similar as they get further apart in time.

The matrix correlation between Mahalanobis distances and time distances is 0.964. Randomization shows that this is also significantly large at about the 2% level.

4.6 Computational methods

The Mahalanobis distance measure requires the calculation of the inverse of a covariance matrix. This can be done using Algorithm 9 of Nash (1979). This is the only complication with the distance measures that have been discussed.

Mantel's test is fairly straightforward to program but a procedure is required for generating a random permutation of n numbers. Algorithm P given by Knuth (1981, p. 139) can be used to do this.

4.7 Further reading

There are a number of aspects of the measurement of distances that have not been covered in this chapter. For example, variables may be binary, indicating presence or absence (1–0) or there may be a mixture of different types of variable that have to be used to measure distances. These matters are discussed by Constandse-Westermann (1972), Gordon (1981) and Romesburg (1984).

References

Constandse-Westerman, T.S. (1972) *Coefficients of Biological Distance.* Anthropological Publications, Oosterhourt, N.B., The Netherlands.

Gordon, A.D. (1981) *Classification.* Chapman and Hall, London.

Higham, C.F.W., Kijngam, A., and Manly, B.F.J. (1980) Analysis of prehistoric canid remains from Thailand. *Journal of Archaeological Science* 7, 149–65.

Knuth, D.E. (1981) *The Art of Computer Programming.* Volume 2, *Seminumerical Algorithms.* Addison-Wesley, Reading, Massachusetts.

Mahalanobis, P.C. (1948) Historical note on the D^3-statistic. *Sankhya* **9**, 237.

Mantel, N. (1967) The detection of disease clustering and a generalized regression approach. *Cancer Research* **27**, 209–20.

Mielke, P.W. (1978) Classification and appropriate inferences for Mantel and Valand's nonparametric multivariate analysis technique. *Biometrics* **34**, 272–82.

Nash, J.C. (1979) *Compact Numerical Methods for Computers.* Adam Hilger, Bristol.

Penrose, L.W. (1953) Distance, size and shape. *Annals of Eugenics* **18**, 337–43.

Romesburg, H.C. (1984) *Cluster Analysis for Researchers.* Lifetime Learning Publications, Belmont, California.

CHAPTER FIVE

Principal component
analysis

5.1 Definition of principal components

The technique of principal component analysis was first described by Karl Pearson (1901). He apparently believed that this was the correct solution to some of the problems that were of interest to biometricians at that time, although he did not propose a practical method of calculation for more than two or three variables. A description of practical computing methods came much later from Hotelling (1933). Even then the calculations were extremely daunting for more than a few variables since they had to be done by hand. It was not until electronic computers became widely available that the technique achieved widespread use.

Principal component analysis is one of the simplest of the multivariate methods that will be described in this book. The object of the analysis is to take p variables $X_1, X_2, \ldots, X_p$ and find combinations of these to produce indices $Z_1, Z_2, \ldots, Z_p$ that are uncorrelated. The lack of correlation is a useful property because it means that the indices are measuring different 'dimensions' in the data. However, the indices are also ordered so that Z_1 displays the largest amount of variation, Z_2 displays the second largest amount of variation, and so on. That is, $\text{var}(Z_1) \geqslant \text{var}(Z_2) \geqslant \ldots \geqslant \text{var}(Z_p)$, where $\text{var}(Z_i)$ denotes the variance of Z_i in the data set being considered. The Z_i are called the principal components. When doing a principal component analysis there is always the hope that the variances of most of the indices will be so low as to be negligible. In that case the variation in the data set can be adequately described by the few Z variables with variances that are not negligible. Some degree of economy is then achieved since the variation in the p original X variables is accounted for by a smaller number of Z variables.

It must be stressed that a principal component analysis does not always work in the sense that a large number of original variables are reduced to a small number of transformed variables. Indeed, if the original variables are uncorrelated then the analysis does absolutely nothing. The best results are obtained when the original variables are very highly correlated, positively or negatively. If that is the case then it is quite conceivable that 20 or 30 original variables can be adequately represented by two or three principal components. If this desirable state of affairs does occur then the important principal components will be of some interest as measures of underlying 'dimensions' in the data. However, it will also be of value to know that there is a good deal of redundancy in the original variables, with most of them measuring similar things.

Before launching into a description of the calculations involved in a principal component analysis it may be of some value to look briefly at the outcome of the analysis when it is applied to the data in Table 1.1 on five body measurements of 49 female sparrows. Details of the analysis are given in Example 5.1. In this case the five measurements are quite highly correlated, as shown in Table 5.1. This is therefore good material for the analysis in question. It turns out, as we shall see, that the first principal component has a variance of 3.62 whereas the other components all have variances very much less than this (0.53, 0.39, 0.30 and 0.16). This means that the first principal component is by far the most important of the five components for representing the variation in the measurements of the 49 birds. The first component is calculated to be

$$Z_1 = 0.45X_1 + 0.46X_2 + 0.45X_3 + 0.47X_4 + 0.40X_5,$$

where $X_1, X_2, \ldots, X_5$ represent here the measurements in Table 1.1

Table 5.1 Correlations between the five body measurements of female sparrows calculated from the data of Table 1.1.

Variable	X_1	X_2	X_3	X_4	X_5
X_1, total length	1.000				
X_2, alar extent	0.735	1.000			
X_3, length of beak & head	0.662	0.674	1.000		
X_4, length of humerus	0.645	0.769	0.763	1.000	
X_5, length of keel of sternum	0.605	0.529	0.526	0.607	1.000

after they have been standardized to have zero means and unit standard deviations. Clearly Z_1 is essentially just an average of the standardized body measurements. It can be thought of as a simple index of size. The analysis given in Example 5.1 leads to the conclusion that most of the differences between the 49 birds are a matter of size (rather than shape).

5.2 Procedure for a principal component analysis

A principal component analysis starts with data on p variables for n individuals, as indicated in Table 5.2. The first principal component is then the linear combination of the variables $X_1, X_2, \ldots, X_p$,

$$Z_1 = a_{11}X_1 + a_{12}X_2 + \cdots + a_{1p}X_p$$

that varies as much as possible for the individuals, subject to the condition that

$$a_{11}^2 + a_{12}^2 + \cdots + a_{1p}^2 = 1.$$

Thus the variance of Z_1, $\text{var}(Z_1)$, is as large as possible given this constraint on the constants a_{1j}. The constraint is introduced because if this is not done then $\text{var}(Z_1)$ can be increased by simply increasing any one of the a_{1j} values. The second principal component,

$$Z_2 = a_{21}X_1 + a_{22}X_2 + \cdots + a_{2p}X_p,$$

is such that $\text{var}(Z_2)$ is as large as possible subject to the constraint that

$$a_{21}^2 + a_{22}^2 + \cdots + a_{2p}^2 = 1,$$

Table 5.2 The form of data for a principal component analysis.

Individual	X_1	X_2	$\cdots$	X_p
1	x_{11}	x_{12}	$\cdots$	x_{1p}
2	x_{21}	x_{22}	$\cdots$	x_{2p}
$\vdots$	$\vdots$	$\vdots$		$\vdots$
n	x_{n1}	x_{n2}	$\cdots$	x_{np}

and also to the condition that Z_1 and Z_2 are uncorrelated. The third principal component,

$$Z_3 = a_{31}X_1 + a_{32}X_2 + \cdots + a_{3p}X_p,$$

is such that var(Z_3) is as large as possible subject to the constraint that

$$a_{31}^2 + a_{32}^2 + \cdots + a_{3p}^2 = 1,$$

and also that Z_3 is uncorrelated with Z_2 and Z_1. Further principal components are defined by continuing in the same way. If there are p variables then there can be up to p principal components.

In order to use the results of a principal component analysis it is not necessary to know how the equations for the principal components are derived. However, it is useful to understand the nature of the equations themselves. In fact a principal component analysis just involves finding the eigenvalues of the sample covariance matrix.

The calculation of the sample covariance matrix has been described in Chapter 2. The important equations are (2.2), (2.3) and (2.7). The matrix is symmetric and has the form

$$\mathbf{C} = \begin{bmatrix} c_{11} & c_{13} & \cdots & c_{1p} \\ c_{21} & c_{23} & \cdots & c_{2p} \\ \vdots & \vdots & & \vdots \\ c_{p1} & c_{p3} & \cdots & c_{pp} \end{bmatrix},$$

where the diagonal element c_{ii} is the variance of X_i and c_{ij} is the covariance of variables X_i and X_j.

The variances of the principal components are the eigenvalues of the matrix $\mathbf{C}$. There are p of these, some of which may be zero. Negative eigenvalues are not possible for a covariance matrix. Assuming that the eigenvalues are ordered as $\lambda_1 \geqslant \lambda_2 \geqslant \cdots \geqslant \lambda_p \geqslant 0$, then λ_i corresponds to the ith principal component

$$Z_i = a_{i1}X_1 + a_{i2}X_2 + \cdots + a_{ip}X_p.$$

In particular, var$(Z_i) = \lambda_i$ and the constants $a_{i1}, a_{i2}, \ldots, a_{ip}$ are the elements of the corresponding eigenvector.

An important property of the eigenvalues is that they add up to the sum of the diagonal elements (the trace) of **C**. That is

$$\lambda_1 + \lambda_2 + \cdots + \lambda_p = c_{11} + c_{22} + \cdots + c_{pp}.$$

Since c_{ii} is the variance of X_i and λ_i is the variance of Z_i, this means that the sum of the variances of the principal components is equal to the sum of the variances of the original variables. Therefore, in a sense, the principal components account for all of the variation in the original data.

In order to avoid one variable having an undue influence on the principal components it is usual to code the variables $X_1, X_2, \ldots, X_p$ to have means of zero and variances of one at the start of an analysis. The matrix **C** then takes the form

$$\mathbf{C} = \begin{bmatrix} 1 & c_{12} & \cdots & c_{1p} \\ c_{21} & 1 & \cdots & c_{2p} \\ \vdots & \vdots & & \vdots \\ c_{p1} & c_{p2} & \cdots & 1 \end{bmatrix}$$

where $c_{ij} = c_{ji}$ is the correlation between X_i and X_j. In other words, the principal component analysis is carried out on the correlation matrix. In that case, the sum of the diagonal terms, and hence the sum of the eigenvalues, is equal to p, the number of variables.

The steps in a principal component analysis can now be stated:

1. Start by coding the variables $X_1, X_2, \ldots, X_p$ to have zero means and unit variances. This is usual, but is omitted in some cases.
2. Calculate the covariance matrix **C**. This is a correlation matrix if step 1 has been done.
3. Find the eigenvalues $\lambda_1, \lambda_2, \ldots, \lambda_p$ and the corresponding eigenvectors $\mathbf{a}_1, \mathbf{a}_2, \ldots, \mathbf{a}_p$. The coefficients of the ith principal component are then given by $\mathbf{a}_i$ while λ_i is its variance.
4. Discard any components that only account for a small proportion of the variation in the data. For example, starting with 20 variables it might be found that the first three components account for 90% of the total variance. On this basis the other 17 components may reasonably be ignored.

Example 5.1 Body measurements of female sparrows

Some mention has already been made of what happens when a principal component analysis is carried out on the data on five body measurements of 49 female sparrows (Table 1.1). It is now worth while to consider the example in more detail.

It is appropriate to begin with step 1 of the four parts of the analysis that have just been described. Standardization of the measurements ensure that they all have equal weight in the analysis. Omitting standardization would mean that the variables X_1 and X_2, which vary most over the 49 birds, would tend to dominate the principal components.

The covariance matrix for the standardized variables is the correlation matrix. This has already been given in lower triangular form in Table 5.1. The eigenvalues of this matrix are found to be 3.616, 0.532, 0.386, 0.302 and 0.164. These add to 5.000, the sum of the diagonal terms in the correlation matrix. The corresponding eigenvectors are shown in Table 5.3, standardized so that the sum of the squares of the coefficients is unity for each one of them. These eigenvectors then provide the coefficients of the principal components.

The eigenvalue for a principal component indicates the variance that it accounts for out of the total variances of 5.000. Thus the first principal component accounts for $(3.616/5.000) 100\% = 72.3\%$, the second for 10.6%, the third for 7.7%, the fourth for 6.0%, and the fifth for 3.3%. Clearly the first component is far more important than the others.

Table 5.3 The eigenvalues and eigenvectors of the correlation matrix for five measurements on 49 female sparrows. The eigenvalues are the variances of the principal components. The eigenvectors give the coefficients of the standardized variables.

Component	Eigenvalue	Eigenvector, coefficient of				
		X_1	X_2	X_3	X_4	X_5
1	3.616	0.452	0.462	0.451	0.471	0.398
2	0.532	−0.051	0.300	0.325	0.185	−0.877
3	0.386	0.691	0.341	−0.455	−0.411	−0.179
4	0.302	−0.420	0.548	−0.606	0.388	0.069
5	0.165	0.374	−0.530	−0.343	0.652	−0.192

Another way of looking at the relative importance of principal components is in terms of their variance in comparison to the variance of the original variables. After standardization the original variables all have variances of 1.0. The first principal component therefore has a variance of 3.616 original variables. However, the second principal component has a variance of only 0.532 of that of one of the original variables. The other principal components account for even less variation.

The first principal component is

$$Z_1 = 0.452X_1 + 0.462X_2 + 0.451X_3 + 0.471X_4 + 0.398X_5,$$

where X_1 to X_5 are standardized variables. This is an index of the size of the sparrows. It seems therefore that about 72.3% of the variation in the data are related to size differences.

The second principal component is

$$Z_2 = -0.051X_1 + 0.300X_2 + 0.325X_3 + 0.185X_4 - 0.877X_5.$$

This appears to be a contrast between variables X_2 (alar extent), X_3 (length of beak and head), and X_4 (length of humerus) on the one hand, and variable X_5 (length of the keel of the sternum) on the other. That is to say, Z_2 will be high if X_2, X_3 and X_4 are high but X_5 is low. On the other hand, Z_2 will be low if X_2, X_3 and X_4 are low but X_5 is high. Hence Z_2 represents a shape difference between the sparrows. The low coefficient of X_1 (total length) means that the value of this variable does not affect Z_2 very much. The principal components Z_3, Z_4 and Z_5 can be interpreted in a similar way. They represent other aspects of shape differences.

The values of the principal components may be useful for further analyses. They are calculated in the obvious way from the standardized variables. Thus for the first bird the original variable values are $x_1 = 156$, $x_2 = 245$, $x_3 = 31.6$, $x_4 = 18.5$ and $x_5 = 20.5$. These standardize to $x_1 = (156 - 157.980)/3.654 = -0.542$, $x_2 = (245 - 241.327)/5.068 = 0.725$, $x_3 = (31.6 - 31.459)/0.795 = 0.177$, $x_4 = (18.5 - 18.469)/0.564 = 0.055$, and $x_5 = (20.5 - 20.827)/0.991 = -0.330$, where in each case the variable mean for the 49 birds has been subtracted and a division has been made by the variable standard deviation for the 49 birds. The value of the first principal component for the first bird is therefore

$$Z_1 = 0.452 \times (-0.542) + 0.462 \times 0.725 + 0.451 \times 0.177$$
$$+ 0.471 \times 0.055 + 0.398 \times (-0.330)$$
$$= 0.064.$$

The second principal component for the bird is

$$Z_2 = -0.051 \times (-0.542) + 0.300 \times 0.725 + 0.325 \times 0.177$$
$$+ 0.185 \times 0.055 - 0.877 \times (-0.330)$$
$$= 0.602.$$

The other principal components can be calculated in a similar way.

The birds being considered were picked up after a severe storm. The first 21 of them recovered while the other 28 died. A question of some interest is therefore whether the survivors and non-survivors show any differences. It has been shown in Example 3.1 that there is no evidence of any differences in mean values. However, in Example 3.2 it has been shown that the survivors seem to have been less variable than the non-survivors. The situation can now be considered in terms of principal components.

The means and standard deviations of the five principal components are shown in Table 5.4 separately for survivors and non-survivors. None of the mean differences between survivors and non-survivors is significant on a t test and none of the standard deviation differences is significant on an F test. However, Levene's test on

Table 5.4 Comparison between survivors and non-survivors in terms of means and standard deviations of principal components.

Principal component	Mean		Standard deviation	
	Survivors	Non-survivors	Survivors	Non-survivors
1	−0.100	0.075	1.506	2.176
2	0.004	−0.003	0.684	0.776
3	−0.140	0.105	0.522	0.677
4	0.073	−0.055	0.563	0.543
5	0.023	−0.017	0.411	0.408

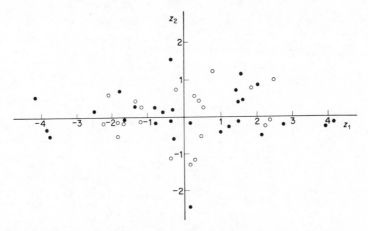

Figure 5.1 Plot of 49 female sparrows against values for the first two principal components, Z_1 and Z_2. (Open circles indicate survivors, closed circles indicate non-survivors.)

deviations from medians (described in Chapter 3) gives a significant difference (just) between the variation of principal component 1 for survivors and non-survivors on a one-sided test at the 5% level. The assumption for the one-sided test is that, if anything, non-survivors were more variable than survivors. The variation is not significantly different for survivors and non-survivors with Levene's test on the other principal components. Since principal component 1 measures overall size, it seems that stabilizing selection may have acted against very large and very small birds.

Figure 5.1 shows a plot of the values of the 49 birds for the first two principal components, which between them account for 82.9% of the variation in the data. The figure shows quite clearly how birds with extreme values for the first principal component failed to survive. Indeed, there is a suggestion that this was true for principal component 2 as well.

Example 5.2 Employment in European countries

As a second example of a principal component analysis, consider the data in Table 1.5 on the percentages of people employed in nine industry sectors in Europe. The correlation matrix for the nine variables is shown in Table 5.5. Overall the values in this matrix are

Table 5.5 The correlation matrix for percentages employed in nine industry groups in 26 countries in Europe, in lower diagonal form, calculated from Table 1.5.

	AGR	MIN	MAN	PS	CON	SER	FIN	SPS	TC
Agriculture	1.000								
Mining	0.036	1.000							
Manufacturing	-0.671	0.445	1.000						
Power supplies	-0.400	0.406	0.385	1.000					
Construction	-0.538	-0.026	0.495	0.060	1.000				
Service industries	-0.737	-0.397	0.204	0.202	0.356	1.000			
Finance	-0.220	-0.443	-0.156	0.110	0.016	0.366	1.000		
Social & Personal Services	-0.747	-0.281	0.154	0.132	0.158	0.572	0.108	1.000	
Transport & communications	-0.565	0.157	0.351	0.375	0.388	0.188	-0.246	0.568	1.000

not particularly high, which indicates that several principal components will be required to account for the variation.

The eigenvalues of the correlation matrix, with percentages of the total of 9.000 in parantheses, are 3.487(38.7%), 2.130(23.6%), 1.099(12.2%), 0.995(11.1%), 0.543(6.0%), 0.383(4.2%), 0.226(2.5%), 0.137(1.5%), and 0.000(0%). The last eigenvalue is exactly zero because the sum of the nine variables being analysed is 100% before standardization. The eigenvector corresponding to this eigenvalue is precisely this sum which, of course, has a zero variance. If any linear combination of the original variables in a principal component analysis is constant then this must of necessity result in a zero eigenvalue.

This example is not as straightforward as the previous one. The first principal component only accounts for about 40% of the variation in the data. Four components are needed to account for 86% of the variation. It is a matter of judgement as to how many components are important. It can be argued that only the first two should be considered since these are the only ones with eigenvalues much more than 1.000. On the other hand, the first four components all have eigenvalues substantially larger than the last five components so perhaps the first four should all be considered. To some extent the choice of the number of components that are important will depend on the use that is going to be made of them. For the present example it will be assumed that a small number of indices are required in order to present the main aspects of differences between the countries. For simplicity only the first two components will be examined further. Between them they account for about 62% of the variation.

The first component is

$$Z_1 = 0.52(\text{AGR}) + 0.00(\text{MIN}) - 0.35(\text{MAN}) - 0.26(\text{PS})$$
$$- 0.33(\text{CON}) - 0.38(\text{SER}) - 0.07(\text{FIN}) - 0.39(\text{SPS})$$
$$- 0.37(\text{TC}),$$

where the abbreviations for variables are stated in Table 1.5. Since the analysis has been done on the correlation matrix, the variables in this equation are the original percentages after they have each been standardized to have a mean of zero and a standard deviation of one. From the coefficients of Z_1 it can be seen that it is primarily a contrast between numbers engaged in agriculture (AGR) and numbers

engaged in manufacturing (MAN), power supplies (PS), construction
(CON), service industries (SER), social and personal services (SPS)
and transport and communications (TC). In making this interpretation the variables with coefficients close to zero are ignored since
they will not affect the value of Z_1 greatly.

The second component is

$$Z_2 = 0.05(AGR) + 0.62(MIN) + 0.36(MAN) + 0.26(PS)$$
$$+ 0.05(CON) - 0.35(SER) - 0.45(FIN) - 0.22(SPS)$$
$$+ 0.20(TC),$$

which primarily contrasts numbers in mining (MIN) and manufacturing (MAN) with numbers in service industries (SER) and finance
(FIN).

Figure 5.2 shows a plot of the 26 countries against their values for
Z_1 and Z_2. The picture is certainly rather meaningful in terms of what
is known about the countries. Most of the Western democracies are
grouped with low values of Z_1 and Z_2. Ireland, Portugal, Spain and
Greece have higher values of Z_1. Turkey and Yugoslavia stand out as
being very high on Z_1. The communist countries other than
Yugoslavia are grouped together with high values for Z_2.

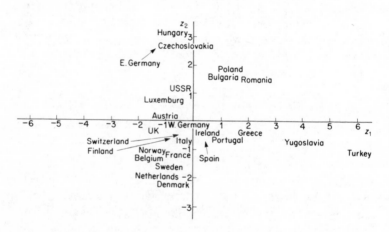

Figure 5.2 European countries plotted against the first two principal components, Z_1 and Z_2 for employment variables.

5.3 Computational methods

Principal component analysis is one of the multivariate techniques that can be programmed reasonably easily on a microcomputer. The eigenvalues and vectors of the covariance or correlation matrix can be determined readily using a standard algorithm such as Algorithm 14 of Nash (1979).

Alternatively, many standard statistical packages will carry out a principal component analysis as one of the multivariate options. In cases where principal component analysis is not specially mentioned it may still be possible to do the calculations using the factor analysis option. More will be said about this in the chapter that follows. Briefly, principal component analysis amounts to a certain type of factor analysis without any rotation of factors.

5.4 Further reading

A short monograph by Daultrey (1976) is recommended for a further discussion of principal component analysis. This is suitable for the general reader although it is aimed particularly at geographers.

References

Daultrey, S. (1976) *Principal Component Analysis*. Concepts and Techniques of Modern Geography, 8, Geo. Abstracts, University of East Anglia, UK.

Hotelling, H. (1933) Analysis of a complex of statistical variables into principal components. *Journal of Educational Psychology* **24**, 417–41, 498–520.

Nash, J.C. (1979) *Compact Numerical Methods for Computers*. Adam Hilger, Bristol.

Pearson, K. (1901) On lines and planes of closest fit to a system of points in space. *Philosophical Magazine* **2**, 557–72.

CHAPTER SIX

Factor analysis

6.1 The factor analysis model

Factor analysis has similar aims to principal component analysis. The basic idea is still that it may be possible to describe a set of p variables $X_1, X_2, \ldots, X_p$ in terms of a smaller number of indices or factors, and hence elucidate the relationship between these variables. There is, however, one important difference: principal component analysis is not based on any particular statistical model, but factor analysis is based on a rather special model.

The early development of factor analysis was due to Charles Spearman. He studied the correlations between test scores of various types and noted that many observed correlations could be accounted for by a simple model for the scores (Spearman, 1904). For example, in one case he obtained the following matrix of correlations for boys in a preparatory school for their scores on tests in Classics (C), French (F), English (E), Mathematics (M), Discrimination of pitch (D), and Music (Mu):-

	C	F	E	M	D	Mu
C	1.00	0.83	0.78	0.70	0.66	0.63
F	0.83	1.00	0.67	0.67	0.65	0.57
E	0.78	0.67	1.00	0.64	0.54	0.51
M	0.70	0.67	0.64	1.00	0.45	0.51
D	0.66	0.65	0.54	0.45	1.00	0.40
Mu	0.63	0.57	0.51	0.51	0.40	1.00

He noted out that this matrix has the interesting property that any two rows are almost proportional if the diagonals are ignored. Thus for rows C and E there are ratios:

$$\frac{0.83}{0.67} \simeq \frac{0.70}{0.64} \simeq \frac{0.66}{0.54} \simeq \frac{0.63}{0.51} \simeq 1.2.$$

Spearman proposed the idea that the six test scores are all of the form

$$X_i = a_i F + e_i,$$

where X_i is the ith standardized score with a mean of zero and a standard deviation of one, a_i is a constant, F is a 'factor' value, which has mean of zero and standard deviation of one for individuals as a whole, and e_i is the part of X_i that is specific to the ith test only. He showed that a constant ratio between rows of a correlation matrix follows as a consequence of these assumptions and that therefore this is a plausible model for the data.

Apart from the constant correlation ratios it also follows that the variance of X_i is given by

$$\begin{aligned}
\mathrm{var}(X_i) &= \mathrm{var}(a_i F + e_i) \\
&= \mathrm{var}(a_i F) + \mathrm{var}(e_i) \\
&= a_i^2 \mathrm{var}(F) + \mathrm{var}(e_i) \\
&= a_i^2 + \mathrm{var}(e_i),
\end{aligned}$$

since a_i is a constant, F and e_i are independent, and the variance of F is assumed to be unity. But $\mathrm{var}(X_i)$ is also unity, so that

$$1 = a_i^2 + \mathrm{var}(e_i).$$

Hence the constant a_i, which is called the *factor loading*, is such that its square is the proportion of the variance of X_i that is accounted for by the factor.

On the basis of his work Spearman formulated his two-factor theory of mental tests: each test result is made up of two parts, one that is common to all tests ('general intelligence'), and another that is specific to the test. Later this theory was modified to allow for each test result to consist of a part due to several common factors plus a part specific to the test. This gives the general factor analysis model

$$X_i = a_{i1} F_1 + a_{i2} F_2 + \cdots + a_{im} F_m + e_i,$$

where X_i is the ith test score with mean zero and unit variance; $a_{i1}, a_{i2}, \ldots, a_{im}$ are the *factor loadings* for the ith test; F_1, F_2, F_m are m uncorrelated *common factors*, each with mean zero and unit variance;

and e_i is a factor specific only to the ith test, which is uncorrelated with any of the common factors and has mean zero.

With this model

$$\operatorname{var}(X_i) = 1 = a_{i1}^2 \operatorname{var}(F_1) + a_{i2}^2 \operatorname{var}(F_2) + \cdots + a_{im}^2 \operatorname{var}(F_m) + \operatorname{var}(e_i)$$
$$= a_{i1}^2 + a_{i2}^2 \cdots + a_{im}^2 + \operatorname{var}(e_i),$$

where $a_{i1}^2 + a_{i2}^2 + \cdots + a_{im}^2$ is called the *communality* of X_i (the part of its variance that is related to the common factors) while $\operatorname{var}(e_i)$ is called the *specificity* of X_i (the part of its variance that is unrelated to the common factors). It can also be established that the correlation between X_i and X_j is

$$r_{ij} = a_{i1} a_{j1} + a_{i2} a_{j2} + \cdots + a_{im} a_{jm}.$$

Hence two test scores can only be highly correlated if they have high loadings on the same factors. Furthermore, $-1 \leqslant a_{ij} \leqslant +1$ since the communality cannot exceed one.

6.2 Procedure for a factor analysis

The data for a factor analysis have the same form as for a principal component analysis. That is, there are p variables with values for these for n individuals, as shown in Table 5.2.

There are three stages to a factor analysis. To begin with, provisional factor loadings a_{ij} are determined. One way to do this is to do a principal component analysis and neglect all of the principal components after the first m, which are themselves taken to be the m factors. The factors found in this way are then uncorrelated with each other and are also uncorrelated with the specific factors. However, the specific factors are not uncorrelated with each other, which means that one of the assumptions of the factor analysis model does not hold. However, this will probably not matter much providing that the communalities are high.

Whatever way the provisional factor loadings are determined, it is possible to show they are not unique. If $F_1, F_2, \ldots, F_m$ are the provisional factors, then linear combinations of these of the form

$$F_1' = d_{11} F_1 + d_{12} F_2 + \cdots + d_{1m} F_m$$
$$F_2' = d_{21} F_1 + d_{22} F_2 + \cdots + d_{2m} F_m$$
$$\vdots$$
$$F_m' = d_{m1} F_1 + d_{m2} F_2 + \cdots + d_{mm} F_m$$

can be constructed that are uncorrelated and 'explain' the data just as well. There are an infinite number of alternative solutions for the factor analysis model, and this leads to the second stage in the analysis, which is called *factor rotation*. Thus the provisional factors are transformed in order to find new factors that are easier to interpret. To 'rotate' in this context means essentially to choose the d_{ij} values in the above equations.

The last stage of an analysis involves calculating the *factor scores*. These are the values of the factors $F_1, F_2, \ldots, F_m$ for each of the individuals.

Generally the number of factors (m) is up to the factor analyst, although it may sometimes be suggested by the nature of the data. When a principal component analysis is used to find a provisional solution, a rough 'rule of thumb' is to choose m equal to the number of eigenvalues greater than unity for the correlation matrix of the test scores. The logic here is the same as was explained in the previous chapter: a factor associated with an eigenvalue of less than unity 'explains' less variation in the overall data than one of the original test scores. In general, increasing m will increase the communalities of variables. However, communalities are not changed by factor rotation.

Factor rotation can be *orthogonal* or *oblique*. With orthogonal rotation the new factors are uncorrelated, like the old factors. With oblique rotation the new factors are correlated. Whichever type of rotation is used, it is desirable that the factor loadings for the new factors should be either close to zero or very different from zero. A near zero a_{ij} means that X_i is not strongly related to the factor F_j. A large (positive or negative) value of a_{ij} means that X_i is determined by F_j to a large extent. If each test score is strongly related to some factors, but not at all related to the others, then this makes the factors easier to identify than would otherwise be the case.

One method of orthogonal factor rotation that is often used is called varimax rotation. This is based on the assumption that the interpretability of factor j can be measured by the variance of the square of its factor loadings, i.e., the variance of $a_{1j}^2, a_{2j}^2, \ldots, a_{pj}^2$. If this variance is large then the a_{ij}^2 values tend to be either close to zero or close to unity. Varimax rotation therefore maximizes the sum of these variances for all the factors. H.F. Kaiser first suggested this approach. Later he modified it slightly by normalizing the factor loadings before maximizing the variances of their squares, since this appears to give improved results (Kaiser, 1958). Varimax rotation can therefore be carried out with or without Kaiser normalization. Numerous other methods of orthogonal rotation have been proposed. However, varimax is recommended as the standard approach.

Sometimes factors analysts are prepared to give up the idea of the factors being uncorrelated in order that the factor loadings should be as simple as possible. An oblique rotation may then give a better solution than an orthogonal one. Again, there are numerous methods available to do the oblique rotation.

Various methods have also been suggested for calculating the factor scores for individuals. A method for use with factor analysis based on principal components is described in the next section. Two other more general methods are estimation by regression and Bartlett's method. See Harman (1976, Chapter 16) for more details.

6.3 Principal component factor analysis

It has been remarked above that one way to do a factor analysis is to begin with a principal component analysis and use the first few principal components as unrotated factors. This has the virtue of simplicity although since the specific factors $e_1, e_2, \ldots, e_p$ are correlated the factor analysis model is not quite correct. Experienced factor analysts often do a principal component factor analysis first and then use other approaches.

The method for finding the unrotated factors is as follows. With p variables there will be the same number of principal components, these being of the form

$$\left.\begin{aligned}
Z_1 &= b_{11}X_1 + b_{12}X_2 + \cdots + b_{1p}X_p \\
Z_2 &= b_{21}X_1 + b_{22}X_2 + \cdots + b_{2p}X_p \\
&\vdots \\
Z_p &= b_{p1}X_1 + b_{p2}X_2 + \cdots + b_{pp}X_p
\end{aligned}\right\} \tag{6.1}$$

where the b_{ij} values are given by the eigenvectors of the correlation matrix. This transformation from X values to Z values is orthogonal, so that the inverse relationship is simply

$$X_1 = b_{11}Z_1 + b_{21}Z_2 + \cdots + b_{p1}Z_p$$
$$X_2 = b_{12}Z_1 + b_{22}Z_2 + \cdots + b_{p2}Z_p$$
$$\vdots$$
$$X_p = b_{1p}Z_1 + b_{2p}Z_2 + \cdots + b_{pp}Z_p.$$

For a factor analysis only m of the principal components are retained, so the last equations become

$$X_1 = b_{11}Z_1 + b_{21}Z_2 + \cdots + b_{m1}Z_m + e_1$$
$$X_2 = b_{12}Z_1 + b_{22}Z_2 + \cdots + b_{m2}Z_m + e_2$$
$$\vdots$$
$$X_p = b_{1p}Z_1 + b_{2p}Z_2 + \cdots + b_{mp}Z_m + e_p$$

All that needs to be done now is to scale the principal components $Z_1, Z_2, \ldots, Z_m$ to have unit variances and hence make them into proper factors. To do this Z_i must be divided by its standard deviation, which is $\sqrt{\lambda_i}$, the square root of the corresponding eigenvalue in the correlation matrix. The equations then become

$$X_1 = \sqrt{\lambda_1}b_{11}F_1 + \sqrt{\lambda_2}b_{21}F_2 + \cdots + \sqrt{\lambda_m}b_{m1}Z_m + e_1$$
$$X_2 = \sqrt{\lambda_1}b_{12}F_1 + \sqrt{\lambda_2}b_{22}F_2 + \cdots + \sqrt{\lambda_m}b_{m2}Z_m + e_2$$
$$\vdots$$
$$X_p = \sqrt{\lambda_1}b_{1p}F_1 + \sqrt{\lambda_2}b_{2p}F_2 + \cdots + \sqrt{\lambda_m}b_{mp}Z_m + e_p$$

where $F_i = Z_i/\sqrt{\lambda_i}$. The unrotated factor model is then

$$\left.\begin{aligned}
X_1 &= a_{11}F_1 + a_{12}F_2 + \cdots + a_{1m}F_m + e_1 \\
X_2 &= a_{21}F_1 + a_{22}F_2 + \cdots + a_{2m}F_m + e_2 \\
&\vdots \\
X_p &= a_{p1}F_1 + a_{p2}F_2 + \cdots + a_{pm}F_m + e_p
\end{aligned}\right\} \tag{6.2}$$

where $a_{ij} = \sqrt{\lambda_j}b_{ji}$.

After rotation (say to varimax loadings) a new solution has the form

$$X_1 = g_{11}F_1^* + g_{12}F_2^* + \cdots + g_{1m}F_m^* + e_1$$
$$X_2 = g_{21}F_1^* + g_{22}F_2^* + \cdots + g_{2m}F_m^* + e_2$$
$$\vdots \tag{6.3}$$
$$X_p = g_{p1}F_1^* + g_{p2}F_2^* + \cdots + g_{pm}F_m^* + e_p$$

where F_i^* represents the new ith factor. The original factors F_i could be expressed exactly as linear combinations of the X variables by scaling equations (6.1). The rotated factors can also still be expressed exactly as linear combinations of the X variables, the relationship being given in matrix form as

$$\mathbf{F}^* = (\mathbf{G}'\mathbf{G})^{-1}\mathbf{G}'\mathbf{X} \qquad (6.4)$$

(Harman, 1976, p. 367), where $(\mathbf{F}^*)' = (F_1^*, F_1^*, \ldots, F_m^*)$, $\mathbf{X}' = (X_1, X_2, \ldots, X_p)$, and $\mathbf{G}$ is the $p \times m$ matrix of factor loadings given in equation (6.3).

6.4 Using a factor analysis program to do principal component analysis

Since many computer programs for factor analysis allow the option of using principal components as initial factors, it is possible to use the programs to do principal component analysis. All that has to be done is to extract the same number of factors as variables and not do any rotation. The factor loadings will then be as given by equations (6.2) with $m = p$ and $e_1 = e_2 = \cdots = e_m = 0$. The principal components are given by equations (6.1) with $b_{ij} = a_{ji}/\sqrt{\lambda_i}$, where λ_i is the ith eigenvalue.

Example 6.1 Employment in European countries

In Example 5.2 a principal component analysis was carried out on the data on the percentages of people employed in nine industry groups in 26 countries in Europe (Table 1.5). It is of some interest to continue the examination of these data using a factor analysis model.

The correlation matrix for the nine percentage variables is given in Table 5.5. The eigenvalues and eigenvectors are shown in Table 6.1. There are three eigenvalues greater than unity so the 'rule of thumb' suggests that three factors should be considered. However, the fourth eigenvalue is almost equal to the third so that either two or four factors can also reasonably be allowed. To begin with, the four-factor solution will be considered.

The eigenvectors in Table 6.1 give the coefficients b_{ij} of the set of equations (6.1). These are changed into factor loadings for four factors as indicated in equations (6.2) to give the factor model:

$$X_1 = \underline{0.98}F_1 + 0.08F_2 - 0.05F_3 + 0.03F_4 \qquad (0.97)$$

$$X_2 = 0.00F_1 + \underline{0.90}F_2 + 0.21F_3 + 0.06F_4 \qquad (0.86)$$

Table 6.1 Eigenvalues and vectors for the European employment data.

		Eigenvector, coefficient of								
	Eigenvalue	X_1	X_2	X_3	X_4	X_5	X_6	X_7	X_8	X_9
1	3.487	0.524	0.001	-0.348	-0.256	-0.325	-0.379	-0.074	-0.387	-0.367
2	2.130	0.054	0.618	0.355	0.261	0.051	-0.350	-0.454	-0.222	0.203
3	1.099	-0.049	0.201	0.151	0.561	-0.153	0.115	0.587	-0.312	-0.378
4	0.995	0.029	0.064	-0.346	0.393	-0.668	-0.050	-0.052	0.412	0.314
5	0.543	0.213	-0.164	-0.385	0.295	0.472	-0.283	0.280	-0.220	0.513
6	0.383	-0.153	0.101	0.289	-0.357	-0.130	-0.615	0.526	0.263	0.124
7	0.226	0.021	-0.726	0.479	0.256	-0.211	0.229	-0.188	-0.191	0.068
8	0.137	0.008	0.088	0.126	-0.341	0.356	0.388	0.174	-0.506	0.545
9	0	-0.806	-0.049	-0.366	-0.019	-0.083	-0.238	-0.145	-0.351	-0.072

$$X_3 = -\underline{0.65}F_1 + \underline{0.52}F_2 + 0.16F_3 - 0.35F_4 \quad (0.83)$$

$$X_4 = -0.48F_1 + 0.38F_2 + \underline{0.59}F_3 + 0.39F_4 \quad (0.87)$$

$$X_5 = -\underline{0.61}F_1 + 0.08F_2 - 0.16F_3 - \underline{0.67}F_4 \quad (0.84)$$

$$X_6 = -\underline{0.71}F_1 - \underline{0.51}F_2 + 0.12F_3 - 0.05F_4 \quad (0.78)$$

$$X_7 = -0.14F_1 - \underline{0.66}F_2 + \underline{0.62}F_3 - 0.05F_4 \quad (0.84)$$

$$X_8 = -\underline{0.72}F_1 - 0.32F_2 - 0.33F_3 + 0.41F_4 \quad (0.90)$$

$$X_9 = -\underline{0.69}F_1 + 0.30F_2 - 0.39F_3 + 0.31F_4 \quad (0.81)$$

The values in parentheses are the communalities. For example, the communality for variable X_1 is $0.98^2 + 0.08^2 + (-0.05)^2 + 0.03^2 = 0.97$, apart from rounding errors. It can be seen that the communalities are fairly high. That is to say, most of the variance for the variables X_1 to X_9 is accounted for by the four common factors.

Factor loadings that are greater than 0.50 (ignoring the sign) are underlined in the above equations. These large and moderate loadings indicate how the variables are related to the factors. It can be seen that X_1 is almost entirely accounted for by factor 1 alone, X_2 is accounted for mainly by factor 2, X_3 is accounted for by factor 1 and factor 2, etc. An undesirable property of this choice of factors is that four of the nine X variables (X_3, X_5, X_6, X_7) are related strongly to two of the factors. This suggests that a rotation may provide simpler factors.

A varimax rotation with Kaiser normalization was carried out. This produced the following model:

$$X_1 = \quad \underline{0.68}F_1 - 0.27F_2 - 0.31F_3 + \underline{0.57}F_4$$

$$X_2 = \quad 0.22F_1 + \underline{0.70}F_2 - \underline{0.55}F_3 - 0.13F_4$$

$$X_3 = \quad 0.13F_1 + 0.49F_2 - 0.12F_3 - \underline{0.75}F_4$$

$$X_4 = -0.23F_1 + \underline{0.89}F_2 - 0.16F_3 - 0.02F_4$$

$$X_5 = -0.16F_1 - 0.11F_2 + 0.03F_3 - \underline{0.90}F_4$$

$$X_6 = -0.53F_1 - 0.03F_2 + 0.62F_3 - 0.33F_4$$

$$X_7 = -0.07F_1 + 0.03F_2 + 0.91F_3 + 0.05F_4$$

$$X_8 = -0.93F_1 - 0.05F_2 + 0.17F_3 - 0.04F_4$$

$$X_9 = -0.77F_1 + 0.23F_2 - 0.33F_3 - 0.23F_4$$

The communalities are unchanged (apart from rounding errors) and the factors are still uncorrelated. However, this is a slightly better solution than the previous one since only X_1, X_2 and X_6 are now appreciably dependent on more than one factor.

At this stage it is usual to try to put labels on factors. It is fair to say that this often requires a degree of inventiveness and imagination! In the present case it is not too difficult.

Factor 1 has a high positive loading for X_1 (agriculture) and high negative loadings for X_6 (service industries), X_8 (social and personal services) and X_9 (transport and communications). It therefore measures the extent to which people are employed in agriculture rather than services, and communications. It can be labelled 'emphasis on agriculture and a lack of service industries'.

Factor 2 has high positive loadings for X_2 (mining) and X_4 (power supplies). This can be labelled 'emphasis on mining and power supplies'.

Factor 3 has high positive loadings on X_6 (service industries) and X_7 (finance) and a high negative loading on X_2 (mining). This can be labelled 'emphasis on financial and service industries rather than mining'.

Finally, factor 4 has high negative loadings on X_3 (manufacturing) and X_5 (construction) and a high positive loading on X_1 (agriculture). 'Lack of industrialization' seems to be a fair label in this case.

The $\mathbf{G}$ matrix of equations (6.3) and (6.4) is given by the factor loadings above. For example, $g_{11} = 0.68$ and $g_{12} = -0.27$, to two decimal places. Carrying out the matrix multiplications and inversion of equation (6.4) produces the equations

$$F_1^* = 0.176X_1 + 0.127X_2 + 0.147X_3 + \cdots - 0.430X_9$$
$$F_2^* = -0.082X_1 + 0.402X_2 + 0.176X_3 + \cdots + 0.014X_9$$
$$F_3^* = -0.122X_1 - 0.203X_2 - 0.025X_3 + \cdots - 0.304X_9$$

Table 6.2 Factor scores for 26 European countries

	FACTOR			
	1 *Agriculture &* *lack of service* *industries*	*2* *Mining &* *power supplies*	*3* *Financial &* *service industries* *& lack of mining*	*4* *Lack of* *industrialization*
Belgium	-0.93	-0.04	0.86	-0.08
Denmark	-1.30	-1.09	0.59	0.44
France	0.02	-0.20	0.98	-0.43
W. Germany	-0.04	0.45	0.45	-0.32
Ireland	-0.32	0.37	0.35	0.82
Italy	0.08	-1.40	-0.07	-1.19
Luxemburg	0.37	0.59	0.18	-1.05
Netherlands	-0.90	-0.59	1.17	-0.24
UK	-0.85	1.23	0.95	0.59
Austria	0.06	0.83	0.68	-0.45
Finland	-0.92	0.47	0.62	0.73
Greece	0.56	-1.12	-0.56	0.42
Norway	-1.77	-0.67	-0.09	0.31
Portugal	0.40	-1.11	-0.07	-0.17
Spain	1.67	-0.64	0.93	-1.67
Sweden	-1.29	-0.38	0.61	0.67
Switzerland	0.68	-0.39	0.98	-1.62
Turkey	1.29	-1.57	-0.85	3.00
Bulgaria	0.26	-0.25	-1.39	-0.34
Czechoslovakia	0.30	1.18	-1.19	-0.63
E. Germany	-0.61	1.70	-1.19	-0.44
Hungary	-0.12	2.37	-1.07	0.42
Poland	0.42	0.26	-1.41	0.06
Romania	1.55	-0.30	-1.11	-0.67
USSR	-0.99	-0.87	-2.06	-0.06
Yugoslavia	2.35	1.17	1.70	1.91

and

$$F_4^* = \quad 0.175X_1 - 0.031X_2 - 0.426X_3 + \cdots + 0.088X_9$$

for estimating factor scores from the data values (after the X variables have been standardized to have zero mean and unit standard deviations). The factor scores obtained from these equations are given in Table 6.2 for the 26 European countries.

From studying the factor scores it can be seen that factor 1 emphasizes the importance of agriculture rather than services and communications in Yugoslavia, Spain, and Romania. The values of factor 2 indicate that countries like Hungary and East Germany have large numbers of people employed in mining and power supplies, with the situation being reversed in countries like Turkey and Italy. Factor 3, on the other hand, is mainly indicating the difference between the communist bloc and the other countries in terms of the numbers employed in finance and service industries. Finally, the values for factor 4 mainly indicate how different Turkey is from the other countries because of its lack of manufacturing and construction workers and relatively large number of agricultural workers.

Most factor analysts would probably continue their analysis of this set of data by trying models with fewer factors and different methods of factor extraction. However, sufficient has been said already to indicate the general approach so the example will be left at this point.

6.5 Options in computer programs

Computer programs for factor analysis often allow many different options and this is likely to be rather confusing for the beginner. For example, BMDP4M, which is the factor analysis option in the Biomedical Computer Programs package (Dixon, 1983), allows four methods for the initial extraction of factors. These are:

1. Principal components, which is the method that was used in the above example. This is recommended for an initial 'look' at data.
2. Maximum likelihood, which is theoretically very good and might well be used as the final method of analysis.
3. 'Little-Jiffy', which is not recommended for beginners.
4. Principal factor analysis, which again is not recommended for beginners.

The program BMDP4M allows even more options when it comes to factor rotation, including no rotation at all. The standard option is varimax rotation. If an oblique rotation is desired, allowing correlated factors, then rotation for simple loading (DQUART in BMDP4M) is recommended.

Finally, there is the question of the number of factors. Most computer programs have an automatic option which can be changed at the user's discretion.

6.6 The value of factor analysis

Factor analysis is something of an art. It is certainly not as objective as most statistical methods. For this reason many statisticians are rather sceptical about its value. For example, Chatfield and Collins (1980, p. 89) list six problems with factor analysis and conclude that 'factor analysis should not be used in most practical situations'. Also, Kendall (1975, p. 59) states that in his opinion 'factor scores are theoretically unmeasurable'.

On the other hand, factor analysis is widely used to analyse data and, no doubt, will continue to be widely used in future. The reason for this is that the technique does seem to be useful for gaining insight into the structure of multivariate data. If it is thought of as a purely descriptive tool then it must take its place as one of the important multivariate methods.

6.7 Computational methods

This chapter has stressed factor analysis based on carrying out a principal component analysis to find initial factors. If this approach is adopted then the main part of the calculation is the finding of the eigenvalues and eigenvectors of the correlation matrix. (See the previous chapter for a suggestion of a suitable algorithm.) Other methods of initial factor extraction are not so straightforward and are probably best done using one of the standard statistical packages.

Varimax has much to recommend it as the standard method for factor rotation. It is quite easy to program since it is done iteratively taking two factors at a time (Harman, 1976, p. 294). Note, however, that Fraenkel (1984) has shown that this may not provide a unique solution and has suggested some modifications to the usual algorithm.

6.8 Further reading

For those about to embark on a factor analysis using a computer program, particularly the Biomedical program BMDP4M, the article by Frane and Hill (1976) should prove of value. The introductory texts by Kim and Mueller (1978a, b) will also be helpful. Those interested in more details should consult one of the specialist texts such as Harman (1976).

References

Chatfield, C. and Collins, A.J. (1980) Introduction to Multivariate Analysis, Chapman and Hall, London.

Dixon, W.J. (Ed.) (1983) BMDP statistical software University of California Press, Berkeley.

Fraenkel, E. (1984) Variants of the varimax rotation method. *Biometrical Journal* 7, 741–8.

Frane, W.J. and Hill, M. (1976) Factor analysis as a tool for data analysis. *Communications in Statistics – Theory and Methods* A5, 487–506.

Harman, H.H. (1976) *Modern Factor Analysis.* University of Chicago Press, Chicago.

Kaiser, H.F. (1958) The varimax criterion for analytic rotation in factor analysis. *Psychology* 23, 187–200.

Kendall, M.G. (1975) *Multivariate Analysis.* Charles Griffin, London.

Kim, J. and Mueller, C.W. (1978a) *Introduction to Factor Analysis.* Sage University Paper Series on Quantitative Applications in the Social Sciences, 07-013. Sage Publications, Beverly Hills.

Kim, J. and Mueller, C.W. (1978b) *Factor Analysis.* Sage University Paper Series on Quantitative Application in the Social Sciences, 07-014. Sage Publications, Beverly Hills.

Spearman, C. (1904) 'General intelligence', objectively determined and measured. *American Journal of Psychology* 15, 201–93.

CHAPTER SEVEN

Discriminant function analysis

7.1 The problem of separating groups

The problem that is addressed with discriminant function analysis is how well it is possible to separate two or more groups of individuals, given measurements for these individuals on several variables. For example, with the data in Table 1.1 on five body measurements of 21 surviving and 28 non-surviving sparrows it is interesting to consider whether it is possible to use the body measurements to separate survivors and non-survivors. Also, for the data shown in Table 1.2 on four dimensions of Egyptian skulls for samples from five time periods it is reasonable to consider whether the measurements can be used to 'age' the skulls.

In the general case there will be m random samples from different groups, of sizes $n_1, n_2, \ldots, n_m$, and values will be available for p variables $X_1, X_2, \ldots, X_p$ for each sample member. Thus the data for a

Table 7.1 The form of data for a discriminant function analysis.

Individual	X_1	X_2	$\cdots$	X_p	
1	x_{111}	x_{112}	$\cdots$	x_{11p}	
2	x_{211}	x_{212}	$\cdots$	x_{21p}	
$\vdots$	$\vdots$	$\vdots$	$\vdots$	$\vdots$	Group 1
n_1	x_{n_111}	x_{n_112}	$\cdots$	x_{n_11p}	
1	x_{121}	x_{122}	$\cdots$	x_{12p}	
2	x_{221}	x_{222}	$\cdots$	x_{22p}	
$\vdots$	$\vdots$	$\vdots$	$\vdots$	$\vdots$	Group 2
n_2	x_{n_221}	x_{n_222}	$\cdots$	x_{n_22p}	
1	x_{1m1}	x_{1m2}	$\cdots$	x_{1mp}	
2	x_{2m1}	x_{2m2}	$\cdots$	x_{2mp}	
$\vdots$	$\vdots$	$\vdots$	$\vdots$	$\vdots$	Group m
n_m	x_{n_mm1}	x_{n_mm2}	$\cdots$	x_{n_mmp}	

discriminant function analysis takes the form shown in Table 7.1.

The data for a discriminant function analysis do not need to be standardized to have zero means and unit variances prior to the start of the analysis, as is usual with principal component and factor analysis. This is because the outcome of a discriminant function analysis is not affected in any important way by the scaling of individual variables.

7.2 Discrimination using Mahalanobis distances

One approach to discrimination is based on Mahalanobis distances, as defined in Section 4.3. The mean vectors for the m samples can be regarded as estimates of the true mean vectors for the groups. The Mahalanobis distances of individuals to group centres can then be calculated and each individual can be allocated to the group that it is closest to. This may or may not be the group that the individual actually came from. The percentage of correct allocations is clearly an indication of how well groups can be separated using the available variables.

This procedure is more precisely defined as follows. Let $\bar{x}_i' = (\bar{x}_{1i}, \bar{x}_{2i}, \ldots, \bar{x}_{pi})'$ denote the vector of mean values for the sample from the ith group, calculated using equations (2.1) and (2.5), and let C_i denote the covariance matrix for the same sample calculated using equations (2.2), (2.3) and (2.7). Also, let C denote the pooled sample covariance matrix determined using equation (4.6). Then the Mahalanobis distance from an observation $x' = (x_1, x_2, \ldots, x_p)'$ to the centre of group i is estimated as

$$D_i^2 = (x - \bar{x}_i)' C^{-1} (x - \bar{x}_i)$$

$$= \sum_{r=1}^{p} \sum_{s=1}^{p} (x_r - \bar{x}_{ri}) c^{rs} (x_s - \bar{x}_{si}). \qquad (7.1)$$

where c^{rs} is the element in the rth row and sth column of C^{-1}. The observation x is allocated to the group for which D_i^2 has the smallest value.

7.3 Canonical discriminant functions

It is sometimes useful to be able to determine functions of the variables $X_1, X_2, \ldots, X_p$ that in some sense separate the m groups as well as is possible. The simplest approach involves taking a linear combination of the X variables

$$Z = a_1 X_1 + a_2 X_2 + \cdots + a_p X_p$$

for this purpose. Groups can be well separated using Z if the mean value changes considerably from group to group, with the values within a group being fairly constant. One way to choose the coefficients $a_1, a_2, \ldots, a_p$ in the index is therefore so as to maximize the F ratio for a one-way analysis of variance. Thus if there are a total of N individuals in all the groups, an analysis of variance on Z values takes the form shown in the following table:

Source of variation	Degrees of freedom	Mean square	F ratio
Between groups	$m - 1$	M_B	M_B/M_W
Within groups	$N - m$	M_W	
	$N - 1$		

Hence a suitable function for separating the groups can be defined as the linear combination for which the F ratio M_B/M_W is as large as possible. This idea was first used by Fisher (1936).

When this approach is used, it turns out that it may be possible to determine several linear combinations for separating groups. In general, the number available is the smaller of p and $m - 1$, say s. They are referred to as *canonical* discriminant functions. The first function

$$Z_1 = a_{11} X_1 + a_{12} X_2 + \cdots + a_{1p} X_p$$

gives the maximum possible F ratio on a one-way analysis of variance for the variation within and between groups. If there is more than one function then the second one

$$Z_2 = a_{21} X_1 + a_{22} X_2 + \cdots + a_{2p} X_p$$

gives the maximum possible F ratio on a one-way analysis of variance subject to the condition that there is no correlation between Z_1 and Z_2 within groups. Further functions are defined in the same way.

Thus the ith canonical discriminant function

$$Z_i = a_{i1}X_1 + a_{i2}X_2 + \cdots + a_{ip}X_p$$

is the linear combination for which the F ratio on an analysis of variance is maximized, subject to Z_i being uncorrelated with $Z_1, Z_2, \ldots,$ and Z_{i-1} within groups.

Finding the coefficients of the canonical discriminant functions turns out to be an eigenvalue problem. The within-sample matrix of sums of squares and cross products, $\mathbf{W}$, has to be calculated using equation (3.12), and also $\mathbf{T}$, the total sample matrix of sums of squares and cross products, calculated using equation (3.11). From these, the between-groups matrix

$$\mathbf{B} = \mathbf{T} - \mathbf{W}$$

can be determined. Next, the eigenvalues and eigenvectors of the matrix $\mathbf{W}^{-1}\mathbf{B}$ have to be found. If the eigenvalues are $\lambda_1 > \lambda_2 > \cdots > \lambda_s$ then λ_i is the ratio of the between-group sum of squares to the within-group sum of squares for the ith linear combination, Z_i, while the elements of the corresponding eigenvector, $\mathbf{a}_i' = (a_{i1}, a_{i2}, \ldots, a_{ip})$, are the coefficients of Z_i.

The canonical discriminant functions $Z_1, Z_2, \ldots, Z_s$ are linear combinations of the original variables chosen in such a way that Z_1 reflects group differences as much as possible; Z_2 captures as much as possible of the group differences not displayed by Z_1; Z_3 captures as much as possible of the group differences not displayed by Z_1 and Z_2; etc. The hope is that the first few functions are sufficient to account for almost all of the important group differences. In particular, if only the first one or two functions are needed for this purpose then a simple graphical representation of the relationship between the various groups is possible by plotting the values of these functions for the sample individuals.

7.4 Tests of significance

Several tests of significance are useful in conjunction with a discriminant function analysis. In particular, the T^2 test of equation (3.5) can be used to test for a significant difference between the mean values for any pair of groups, while the likelihood ratio test of equation (3.10) can be used to test for overall differences between the means of the m groups.

In addition, a test is available to see whether the canonical discriminant function Z_j varies significantly from group to group. Here the test statistic is

$$\phi_j^2 = \left\{ \sum_{i=1}^{m} n_i - 1 - \tfrac{1}{2}(p + m) \right\} \log_e(1 + \lambda_j), \qquad (7.2)$$

which is tested against the chi-squared distribution with $p + m - 2j$ degrees of freedom. A significantly large value for ϕ_j^2 indicates that Z_j does vary from group to group.

Finally, the Mahalanobis distances from observations to their group centres can be examined. These distances should be approximately chi-square variates with p degrees of freedom. If an observation is very significantly far from the centre of its group on the basis of the chi-square distribution then this brings into question whether the observation really comes from this group (see p. 48).

7.5 Assumptions

The methods of discrimination given in Sections 7.2 and 7.3, and the tests of significance just described, are based on the assumption that the within-group covariance matrix is the same for all groups. If this is not true then the analyses may not work very well. In particular, tests of significance are not reliable. Also, tests of significance require the assumption that within groups the data follow multivariate normal distributions. In general it seems that multivariate analyses that assume normality can be upset quite badly if the assumption is not correct. This contrasts with the situation with univariate analyses such as regression analysis and analysis of variance, which are generally quite robust to the assumption of normality.

It should, however, be noted that a failure of one or more assumptions does not necessarily mean that a discriminant function analysis is a waste of time. It may well turn out that excellent discrimination is possible on non-normal populations. The only problem is that it may not be simple to establish the significance of results.

Example 7.1 The storm survival of female sparrows

As a first example, a case will be considered where a discriminant function analysis does not produce useful results. This concerns differences in five morphological variables between the female sparrows that survived and those that did not survive a severe storm. The data are given in Table 1.1. It has been shown already that there are no significant differences between the mean values of survivors and non-survivors (Example 3.1), although the non-survivors may possibly have been more variable (Example 3.2). A principal component analysis has confirmed the test results (Example 5.1).

Given these preliminary analyses it can come as no surprise that insignificant results are found from a discriminant function analysis. With $m = 2$ groups there is only a single canonical discriminant function, which corresponds to the single non-zero eigenvalue $\lambda_1 = 0.033$ of the matrix $\mathbf{W}^{-1}\mathbf{B}$. This is tested for significance by calculating ϕ_1^2 from equation (7.2) as

$$\phi_1^2 = \{21 + 28 - 1 - \tfrac{1}{2}(5 + 2)\} \log_e(1 + 0.033)$$
$$= 1.44,$$

and comparing this with critical values from the chi-squared distribution with $p + m - 2j = 5 + 2 - 2 = 5$ degrees of freedom. Clearly ϕ_1^2 is not significantly large. There is therefore no point in taking the analysis further.

Example 7.2 Comparison of samples of Egyptian skulls

A more promising situation concerns the comparison of the values for four measurements on male Egyptian skulls for five samples ranging in age from the early predynastic period (*circa* 4000 BC) to the Roman period (*circa* AD 150). In this case the data are shown in Table 1.2. It has already been established that the mean values differ significantly from sample to sample (Example 3.3), with the differences tending to increase with the time difference between samples (Example 4.3).

The within-sample matrix of sums of squares and cross products is calculated using equation (3.12) as

$$\mathbf{W} = \begin{bmatrix} 3061.67 & 5.33 & 11.47 & 291.30 \\ 5.33 & 3405.27 & 754.00 & 412.53 \\ 11.47 & 754.00 & 3505.97 & 164.33 \\ 291.30 & 412.53 & 164.33 & 1472.13 \end{bmatrix},$$

while the corresponding total sample matrix is calculated using equation (3.11) as

$$\mathbf{T} = \begin{bmatrix} 3563.89 & -222.81 & -615.16 & 426.73 \\ -222.81 & 3635.17 & 1046.28 & 346.47 \\ -615.16 & 1046.28 & 4309.27 & -16.40 \\ 426.73 & 346.47 & -16.40 & 1533.33 \end{bmatrix}.$$

The between-sample matrix is therefore

$$\mathbf{B} = \mathbf{T} - \mathbf{W} = \begin{bmatrix} 502.83 & -228.15 & -626.63 & 135.43 \\ -228.15 & 229.91 & 292.28 & -66.07 \\ -626.63 & 292.28 & 803.30 & -180.73 \\ 135.43 & -66.07 & -180.73 & 61.20 \end{bmatrix}.$$

The eigenvalues of $\mathbf{W}^{-1}\mathbf{B}$ are found to be $\lambda_1 = 0.437$, $\lambda_2 = 0.035$, $\lambda_3 = 0.015$ and $\lambda_4 = 0.002$. The corresponding canonical discriminant functions are

$$\left.\begin{aligned} Z_1 &= -0.0107X_1 + 0.0040X_2 + 0.0119X_3 - 0.0068X_4 \\ Z_2 &= 0.0031X_1 + 0.0168X_2 - 0.0046X_3 - 0.0022X_4 \\ Z_3 &= -0.0068X_1 + 0.0010X_2 + 0.0000X_3 + 0.0247X_4 \end{aligned}\right\} \quad (7.3)$$

and

$$Z_4 = 0.0126X_1 - 0.0001X_2 + 0.0112X_3 + 0.0054X_4$$

From equation (7.2) the chi-square values for the significance of the components are $\phi_1^2 = 52.40$, with 7 degrees of freedom, $\phi_2^2 = 5.03$ with 5 degrees of freedom, $\phi_3^2 = 2.16$ with 3 degrees of freedom, and $\phi_4^2 = 0.30$ with 1 degree of freedom. Of these, only ϕ_1^2 is significantly large at the 5% level. It can therefore be concluded that the differences between the five samples are adequately described by Z_1 alone.

Table 7.2 Means and standard deviations for the canonical discriminant function Z_1 for five samples of Egyptian skulls.

Sample	Mean	Std. dev.
Early predynastic	− 0.029	0.097
Late predynastic	− 0.043	0.071
12th & 13th dynasties	− 0.099	0.075
Ptolemaic	− 0.143	0.080
Roman	− 0.167	0.095

The X variables in equations (7.3) are the values as shown in Table 1.2 without standardization. They are illustrated in Fig. 1.1 from which it can be seen that large values of Z_1 correspond to skulls that are tall but narrow, with a long jaw and a short nasal height.

The Z_1 values for individuals are calculated in the obvious way. For example, the first individual in the early predynastic sample has $X_1 = 131$ mm, $X_2 = 138$ mm, $X_3 = 89$ mm and $X_4 = 49$ mm. Therefore

$$Z_1 = - 0.0107 \times 131 + 0.0040 \times 138 + 0.0119 \times 89$$
$$- 0.0068 \times 49 = - 0.124$$

is the value of the discriminant function in this case. The means and standard deviations found for the Z_1 values for the five samples are shown in Table 7.2. It can be seen quite clearly that the mean of Z_1 has become lower over time, indicating a trend towards shorter, broader skulls with short jaws but relatively large nasal heights. This is,

Table 7.3 Results obtained when 150 Egyptian skulls are allocated to the groups for which they have the minimum Mahalanobis distance.

Source group	Number allocated to group					
	1	2	3	4	5	Total
1	12	8	4	4	2	30
2	10	8	5	4	3	30
3	4	4	15	2	5	30
4	3	3	7	5	12	30
5	2	4	4	9	11	30

however, very much an average change. If the 150 skulls are allocated to the samples to which they are closest according to the Mahalanobis distance function of equation (7.1) then only a fairly small proportion are allocated to the samples that they really belong to (Table 7.3). Thus although this discriminant function analysis has been successful in pinpointing the changes in skull dimensions over time, it has not produced a satisfactory method for 'ageing' skulls.

Example 7.3 Discriminating between groups of European countries

The data shown in Table 1.5 on employment percentages in nine groups in 26 European countries have already been examined by principal component analysis and by factor analysis (Examples 5.2 and 6.1). Here they will be considered from the point of view of the extent to which it is possible to discriminate between groups of countries on the basis of employment patterns. In particular, three natural groups existed in 1979 when the data were collected. These were: (1) the European Economic Community (EEC) countries at the time of Belgium, Denmark, France, West Germany, Ireland, Italy, Luxemburg, the Netherlands and the United Kingdom; (2) the other western European countries of Austria, Finland, Greece, Norway, Portugal, Spain, Sweden, Switzerland and Turkey; and (3) the eastern European communist countries of Bulgaria, Czechoslovakia, East Germany, Hungary, Poland, Romania, USSR and Yugoslavia. These three groups can be used as a basis for a discriminant function analysis.

The percentages in the nine industry groups add to 100% for each of the 26 countries. This means that any one of the nine percentage variables can be expressed as 100 minus the remaining variables. It is therefore necessary to omit one of the variables from the analysis in order to calculate Mahalanobis distances and canonical discriminant functions. The last variable, the percentage employed in transport and communications, was omitted for the analysis that will now be described.

The number of canonical variables is two in this example, this being the minimum of the number of variables ($p = 8$) and the number of groups minus one ($m - 1 = 2$). These canonical variables are

$$Z_1 = 0.73\text{AGR} + 0.62\text{MIN} + 0.63\text{MAN} - 0.16\text{PS}$$

$$+ 0.80\text{CON} + 1.24\text{SER} + 0.72\text{FIN} + 0.82\text{SPS},$$

and

$$Z_2 = 0.84\text{AGR} + 2.46\text{MIN} + 0.78\text{MAN} + 1.18\text{PS}$$
$$+ 1.17\text{CON} + 0.83\text{SER} + 0.84\text{FIN} + 1.05\text{SPS},$$

the corresponding eigenvalues of $\mathbf{W}^{-1}\mathbf{B}$ being $\lambda_1 = 7.531$ and $\lambda_2 = 1.046$. The corresponding chi-square values from equation (7.2) are $\phi_1^2 = 41.80$, with 9 degrees of freedom, and $\phi_2^2 = 13.96$, with 7 degrees of freedom. The chi-square value for Z_1 is significantly large at the 0.1% level. The chi-square value for Z_2 is not quite significantly large at the 5% level.

From the coefficients in the equation for Z_1 it can be seen that this variable will tend to be large when there are high percentages employed in everything except PS (power supplies). There is a particularly high coefficient for SER (service industries). For Z_2, on the other hand, all the coefficients are positive, with that for MIN (mining) being particularly high.

A plot of the countries against their values for Z_1 and Z_2 is shown in Fig. 7.1. The eastern European communist countries appear on the left-hand side, the non-EEC western European countries in the centre, and the EEC countries on the right-hand side of the figure. It can be clearly seen how most separation occurs with the horizontal

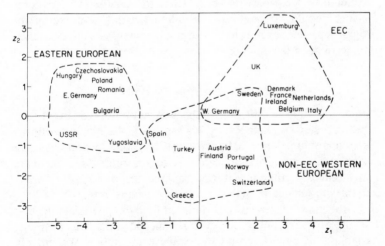

Figure 7.1 Plot of 26 European countries against their values for two canonical discriminant functions.

values of Z_1. As far as values of Z_2 are concerned, it appears that the non-EEC western European countries tend to have lower values than the other two groups. Overall, the degree of separation of the three groups is good. The only 'odd' cases are West Germany, which appears to be more like a non-EEC western European country than an EEC country, and Sweden, which appears to be more like an EEC country than a non-EEC western European country.

The discriminant function analysis has been rather successful in this example. It is possible to separate the three groups of countries on the basis of their employment patterns. Furthermore, the separation using the two canonical discriminant functions is much clearer than the separation shown in Fig. 5.2 (p. 70) for the first two principal components.

7.6 Allowing for prior probabilities of group membership

Computer programs allow many options for varying a discriminant function analysis. One situation is that the probability of membership is inherently different for different groups. For example, if there are two groups it might be that it is known that most individuals fall into group 1 while very few fall into group 2. In that case if an individual is to be allocated to a group it makes sense to bias the allocation procedure in favour of group 1. Thus the process of allocating an individual to the group to which it has the smallest Mahalanobis distance should be modified. To allow for this some computer programs permit prior probabilities of group membership to be taken into account in the analysis.

7.7 Stepwise discriminant function analysis

Another possible modification of the basic analysis involves carrying it out in a stepwise manner. In this case variables are added to the discriminant functions one by one until it is found that adding extra variables does not give significantly better discrimination. There are many different criteria that can be used for deciding on which variables to include in the analysis and which to miss out.

A problem with stepwise discriminant function analysis is the bias that the procedure introduces into significance tests. Given enough variables it is almost certain that some combination of them will produce (significant) discriminant functions by chance alone. If a stepwise analysis is carried out then it is advisable to check its validity by rerunning it several times with a random allocation of individuals to groups to see how significant are the results obtained. For example, with the Egyptian skull data the 150 skulls could be allocated completely at random to five groups of 30, the allocation being made a number of times, and a discriminant function analysis run on each random set of data. Some idea could then be gained of the probability of getting significant results through chance alone.

It should be stressed that this type of randomization to verify a discriminant function analysis is unnecessary for a standard non-stepwise analysis providing there is no reason to suspect the assumptions behind the analysis. It could, however, be informative in cases where the data are clearly not normally distributed within groups or where the within-group covariance matrix is not the same for each group.

7.8 Jackknife classification of individuals

A moment's reflection will suggest that an allocation matrix such as that shown in Table 7.3 must tend to have a bias in favour of allocating individuals to the group that they really come from. After all, the group means are determined from the observations in that group. It is not surprising to find that an observation is closest to the centre of the group where the observation helped to determine that centre.

To overcome this bias, some computer programs carry out what is called a 'jackknife classification' of observations. This involves allocating each individual to its closest group without using that individual to help determine a group centre. In this way any bias in the allocation is avoided. In practice there is often not a great deal of difference between the straightforward classification and the jackknife classification. The jackknife classification usually gives a slightly smaller number of correct allocations.

7.9 Assigning of ungrouped individuals to groups

Some computer programs allow the input of data values for a number of individuals for which the true group is not known. It is then possible to assign these individuals to the group that they are closest to, in the Mahalanobis distance sense, on the assumption that they have to come from one of the m groups that are sampled. Obviously in these cases it will not be known whether the assignment is correct. However, the errors in the allocation of individuals from known groups is an indication of how accurate the assignment process is likely to be. For example, the results shown in Table 7.3 indicate that allocating Egyptian skulls to different time periods using skull dimensions is liable to result in many errors.

7.10 Computational methods

The only complicated part of carrying out an analysis along the lines described in this chapter is finding the eigenvalues and vectors of the matrix $\mathbf{W}^{-1}\mathbf{B}$, which is not symmetric. However, Algorithm 15 of Nash (1979) will do the required calculations if the eigenvalue problem

$$(\mathbf{W}^{-1}\mathbf{B} - \lambda\mathbf{I})\mathbf{a} = \mathbf{0}$$

is written as

$$(\mathbf{B}\mathbf{W}^{-1}\mathbf{B} - \lambda\mathbf{B})\mathbf{a} = \mathbf{0}.$$

The matrix inverse $\mathbf{C}^{-1}$ of equation (7.1) can be found using Algorithm 5 or Algorithm 9 from the same book.

7.11 Further reading

The description 'linear discriminant function analysis' usually means an analysis based on finding the canonical discriminant functions as described in Section 7.3, with individuals being assigned to groups according to smallest Mahalanobis distances.

As mentioned before, an assumption of this approach to discrimination is that the several populations sampled all have the same covariance matrix. If this assumption is not true then it can be shown that it is appropriate to use quadratic discriminant functions instead of linear ones. If the distributions being sampled are also very non-normal then a completely different approach to discrimination may be required. Lachenbruch and Goldstein (1979) and Fatti *et al.* (1982) provide references to the various procedures that have been suggested.

References

Fatti, L.P., Hawkins, D.M. and Raath, E.L. (1982) In *Topics in Applied Multivariate Analysis* (Ed. D.M. Hawkins), pp. 1–71. Cambridge University Press, Cambridge.

Fisher, R.A. (1936) The utilization of multiple measurements in taxonomic problems. *Annals of Eugenics* **7**, 179–88.

Lachenbruch, P.A. and Goldstein, M. (1979) Discriminant analysis. *Biometrics* **35**, 69–85.

Nash, J.C. (1979) *Compact Numerical Methods for Computers.* Adam Hilger, Bristol.

— CHAPTER EIGHT —

Cluster analysis

8.1 Uses of cluster analysis

The problem that cluster analysis is designed to solve is the following one: given a sample of n objects, each of which has a score on p variables, devise a scheme for grouping the objects into classes so that 'similar' ones are in the same class. The method must be completely numerical and the number of classes is not known. This problem is clearly more difficult than the problem for a discriminant function analysis since with discriminant function analysis the groups are known to begin with.

There are many reasons why cluster analysis may be worth while. Firstly, it might be a question of finding the 'true' groups. For example, in psychiatry there has been a great deal of disagreement over the classification of depressed patients, and cluster analysis has been used to define 'objective' groups. Secondly, cluster analysis may be useful for data reduction. For example, a large number of cities can potentially be used as test markets for a new product but it is only feasible to use a few. If cities can be groupd into a small number of groups of similar cities then one member from each group could be used for the test market. On the other hand, if cluster analysis generates unexpected groupings then this might in itself suggest relationships to be investigated.

8.2 Types of cluster analysis

Many algorithms have been proposed for cluster analysis. Here attention will be restricted to those following two particular approaches. Firstly, there are hierarchic techniques which produce a *dendrogram* such as the ones shown in Fig. 8.1. These methods start with the calculation of the distances of each individual to all other individuals. Groups are then formed by a process of agglomeration or division. With agglomeration all objects start by being alone in groups of one. Close groups are then gradually merged until finally all individuals are in a single group. With division all objects start in a single group. This is then split into two groups, the two groups are then split, and so on until all objects are in groups of their own.

The second approach to cluster analysis involves partitioning, with objects being allowed to move in and out of groups at different stages of the analysis. To begin with, some more or less arbitrary group centres are chosen and individuals are allocated to the nearest one. New centres are then calculated where these are at the centres of the individuals in groups. An individual is then moved to a new group if it is closer to that group's centre than it is to the centre of its present group. Groups 'close' together are merged; spread out groups are split, etc. The process continues iteratively until stability is achieved with a predetermined number of groups. Usually a range of values is tried for the final number of groups. The results of a partitioning cluster analysis are considered in Example 8.1.

8.3 Hierarchic methods

As mentioned above, agglomerative hierarchic methods start with a matrix of 'distances' between individuals. All individuals begin alone in groups of one and groups that are 'close' together are merged. (Measures of 'distance' will be discussed later.) There are various ways to define 'close'. The simplest is in terms of *nearest neighbours*. For example, suppose there is the following distance matrix for five objects:

	1	2	3	4	5
1	—				
2	2	—			
3	6	5	—		
4	10	9	4	—	
5	9	8	5	3	—

The calculations are then as shown in the following table. Groups are merged at a given level of distance if one of the individuals in one group is that distance or closer to at least one individual in the second group.

Cluster analysis

Distance	Groups
0	1, 2, 3, 4, 5
2	(1, 2), 3, 4, 5,
3	(1, 2), 3, (4, 5)
4	(1. 2), (3, 4, 5)
5	(1, 2, 3, 4, 5)

At a distance of 0 all five objects are on their own. The distance matrix shows that the smallest distance between two objects is 2, between the first and second objects. Hence at a distance level of 2 there are four groups (1,2), (3), (4) and (5). The next smallest distance between objects is 3, between objects 4 and 5. Hence at a distance of 3 there are three groups (1, 2), (3) and (4, 5). The next smallest distance is 4, between objects 3 and 4. Hence at this level of distance there are two groups (1, 2) and (3, 4, 5). Finally, the next smallest distance is 5, between objects 2 and 3 and between objects 3 and 5. At this level the two groups merge into the single group (1, 2, 3, 4, 5) and the analysis is complete. The dendrogram shown in Fig. 8.1(a) illustrates how agglomeration takes place.

With *furthest neighbour* linkage two groups merge only if the most distant members of the two groups are close enough together. With the example data this work as follows:

Distance	Groups
0	1, 2, 3, 4, 5
2	(1, 2), 3, 4, 5
3	(1, 2), 3, (4, 5)
5	(1, 2), (3, 4, 5)
10	(1, 2, 3, 4, 5)

Object 3 does not join with objects 4 and 5 until distance level 5 since this is the distance to object 3 from the furthest away of objects 4 and 5. The furthest-neighbour dendrogram is shown in Fig. 8.1(b).

With *group average* linkage two groups merge if the average

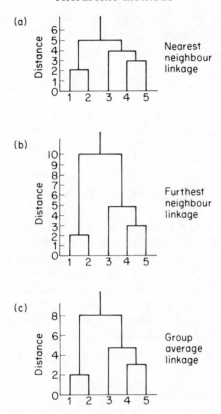

Figure 8.1 Examples of dendrograms from cluster analyses of five objects.

distance between them is small enough. With the example data this gives the following result:

Distance	Groups
0	1, 2, 3, 4, 5
2	(1, 2), 3, 4, 5
3	(1, 2), 3, (4, 5)
4.5	(1, 2), (3, 4, 5)
7.8	(1, 2, 3, 4, 5)

For instance, groups (1, 2) and (3, 4, 5) merge at distance level 7.8 since
this is the average distance from objects 1 and 2 to objects 3, 4 and 5,
the actual distances being:

1–3	6
1–4	10
1–5	9
2–3	5
2–4	9
2–5	8

Mean = 7.8

The dendrogram in this case is shown in Fig. 8.1(c).

Divisive hierarchic methods have been used less often than
agglomerative ones. The objects are all put into one group initially,
and then this is split into two groups by separating off the object that
is furthest on average from the other objects. Individuals from the
main group are then moved to the new group if they are closer to it
than they are to the main group. Further subdivisions occur as the
distance that is allowed between individuals in the same group is
reduced. Eventually all objects are in groups of their own.

8.4 Problems of cluster analysis

It has already been mentioned that there are many algorithms for
cluster analysis. However, there is no generally accepted 'best'
method. Unfortunately, different algorithms do not necessarily
produce the same results on a given set of data. There is usually rather
a large subjective component in the assessment of the results from any
particular method.

A fair test of any algorithm is to take a set of data with a known
group structure and see whether the algorithm is able to reproduce
this structure. It seems to be the case that this test only works in cases
where the groups are very distinct. When there is a considerable
overlap between the initial groups, a cluster analysis may produce a
solution that is quite different from the true situation.

In some cases difficulties will arise because of the shape of clusters.
For example, suppose that there are two variables X_1 and X_2 and
individuals are plotted according to their values for these. Some

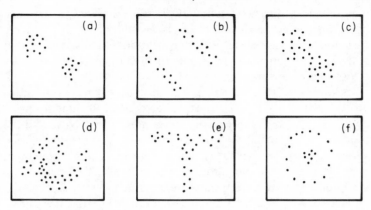

Figure 8.2 Some possible patterns of points with two clusters.

possible patterns of points are illustrated in Fig. 8.2. Case (a) is likely
to be found by any reasonable algorithm, as is case (b). In case (c) some
algorithms might well fail to detect two clusters because of the
intermediate points. Most algorithms would have trouble handling
cases like (d), (e) and (f).

Of course, clusters can only be based on the variables that are given
in the data. Therefore they must be relevant to the classification
wanted. To classify depressed patients there is presumably not much
point in measuring height, weight, or length of arms. A problem here
is that the clusters obtained may be rather sensitive to the particular
choice of variables that is made. A different choice of variables,
apparently equally reasonable, may give rather different clusters.

8.5 Measures of distance

The data for a cluster analysis usually consists of the values of p
variables $X_1, X_2, \ldots, X_p$ for n objects. For hierarchic algorithms these
variable values are then used to produce an array of distances
between the individuals. Measures of distance have already been
discussed in Chapter 4. Here it suffices to say that the Euclidean
distance function

$$d_{ij} = \sqrt{\left\{ \sum_{k=1}^{p} (x_{ik} - x_{jk})^2 \right\}} \tag{8.1}$$

is most frequently used for quantitative variables. Here x_{ik} is the value of variable X_k for individual i and x_{jk} is the value of the same variable for individual j. The geometrical interpretation of the distance d_{ij} from individual i to individual j is illustrated in Figs 4.1 and 4.2 for the cases of two and three variables.

Usually variables are standardized in some way before distances are calculated, so that all p variables are equally important in determining these distances. This can be done by coding so that the means are all zero and the variances are all one. Alternatively, each variable can be coded to have a minimum of zero and a maximum of one. Unfortunately, standardization has the effect of minimizing group differences since if groups are separated well by X_i then the variance of X_i will be large, and indeed it should be large. It would be best to be able to make the variances equal to one within clusters but this is obviously not possible since the whole point of the analysis is to find the clusters.

8.6 Principal component analysis with cluster analysis

Some cluster analysis algorithms begin by doing a principal component analysis to reduce a large number of original variables down to a smaller number of principal components. This can drastically reduce the computing time for the cluster analysis. However, it is known that the results of a cluster analysis can be rather different with and without the initial principal component analysis. Consequently an initial principal component analysis is probably best avoided.

On the other hand, when the first two principal components account for a high percentage of variation in the data a plot of individuals against these two components is certainly a useful way for looking for clusters. For example, Fig. 5.2 (p. 70) shows European countries plotted in this way for principal components based on employment percentages. The countries do seem to group in a meaningful way.

Example 8.1 Clustering of European countries

The data just mentioned on the percentages of people employed in nine industry groups in different countries of Europe (Table 1.5) can be used for a first example of cluster analysis. The analysis should show which countries have similar employment patterns and which countries are different in this respect. It may be recalled from

Table 8.1 Standardized data on the percentages employed in nine industry groups in Europe, as derived from Table 1.5.

Country	AGR	MIN	MAN	PS	CON	SER	FIN	SPS	TC
Belgium	-1.02	-0.36	0.09	-0.02	0.02	1.34	0.78	0.96	0.47
Denmark	-0.64	-1.19	-0.74	0.81	0.08	0.36	0.89	1.78	0.40
France	-0.54	-0.47	0.07	-0.02	0.45	0.84	0.71	0.38	-0.61
West Germany	-0.80	0.05	1.25	-0.02	-0.53	0.32	0.36	0.33	-0.32
Ireland	0.26	-0.26	-0.90	1.04	-0.40	0.84	-0.43	0.11	-0.32
Italy	-0.21	-0.67	0.08	-1.08	1.12	1.12	-0.85	0.01	-0.61
Luxemburg	-0.73	1.90	0.54	-0.29	0.63	1.21	0.21	-0.12	-0.25
Netherlands	-0.82	-1.19	-0.64	0.24	1.05	1.10	1.00	1.24	0.18
UK	-1.06	0.15	0.46	1.31	-0.77	0.86	0.61	1.21	-0.10
Austria	-0.41	-0.16	0.46	1.31	0.51	0.84	0.32	-0.47	0.33
Finland	-0.39	-0.88	-0.16	1.04	-0.46	0.38	0.53	0.63	0.76
Greece	1.43	-0.67	-1.34	-0.82	-0.04	-0.32	-0.57	-1.32	0.11
Norway	-0.65	-0.78	-0.66	-0.29	0.26	0.86	0.25	1.11	2.05
Portugal	0.56	-0.98	-0.36	-0.82	0.14	0.07	-0.46	-0.49	-0.61
Spain	0.24	-0.47	0.21	-0.55	2.03	-0.71	1.60	-1.20	-0.75
Sweden	-0.84	-0.88	-0.16	-0.29	-0.59	0.31	0.71	1.81	0.18
Switzerland	-0.73	-1.09	1.54	-0.29	0.81	0.99	0.46	-0.68	-0.61
Turkey	3.07	-0.57	-2.73	-2.15	-3.26	-1.70	-1.03	-1.19	-2.40
Bulgaria	0.29	0.67	0.76	-0.82	-0.16	-1.08	-1.18	-0.27	0.11
Czechoslovakia	0.17	1.70	1.21	0.78	0.32	-0.82	-1.11	-0.31	0.33
East Germany	-0.96	1.70	2.03	1.04	-0.34	-0.38	-1.00	0.30	1.33
Hungary	0.16	1.90	0.37	2.64	0.02	-0.78	-1.11	-0.41	1.04
Poland	0.77	1.28	-0.19	-0.02	0.14	-1.19	-1.11	-0.57	0.25
Romania	1.00	0.87	0.44	-0.82	0.32	-1.54	-0.96	-1.22	-1.11
USSR	0.29	0.15	-0.17	-0.82	0.63	-1.50	-1.25	0.52	1.98
Yugoslavia	1.90	0.25	-1.46	0.51	-1.98	-1.43	2.60	-2.16	-1.83

Example 7.3 that a sensible grouping into EEC, non-EEC western European countries and eastern European countries existed in 1979, when the data were collected.

An analysis was carried out using the BMDP program 2M (Dixon, 1983), with preassigned options for the various computational methods. Thus the first step in the analysis involved standardizing the nine variables so that each one had a mean of zero and a standard deviation of one. For example, variable 1 is AGR, the percentage employed in agriculture. For the 26 countries being considered this variable has a mean of 19.13 and a standard deviation of 15.55. The data value for Belgium for AGR is 3.3 which standardizes to $(3.3 - 19.13)/15.55 = -1.02$; the data value for Denmark is 9.2 which standardizes to -0.64; and so on. The standardized values are shown in Table 8.1.

The next step in the analysis involved calculating the Euclidean distances between all pairs of countries. This was done using equation (8.1) on the standardized data values. Finally, a dendrogram was formed by the agglomerative, nearest neighbour, hierarchic process described above.

The dendrogram is shown in Fig. 8.3, as output by the BMDP2M computer program. It can be seen that the two closest countries were Sweden and Denmark. These are distance 1.135 apart. The next closest pair of countries are Belgium and France, which are 1.479 apart. Then come Poland and Bulgaria, which are 1.537 apart. Amalgamation ended with Turkey joining the other countries at a distance of 5.019.

Having obtained the dendrogram, we are free to decide how many clusters to take. For example, if six clusters are to be considered then these are found at an amalgamation distance of 2.459. The first cluster is the western nations of Belgium, France, Netherlands, Sweden, Denmark, West Germany, Finland, UK, Austria, Ireland, Switzerland, Norway, Greece, Portugal and Italy. The second cluster is Luxemburg on its own. Then there are the communist countries of USSR, Hungary, Czechoslovakia, East Germany, Romania, Poland and Bulgaria. The last three clusters are Spain, Yugoslavia and Turkey, each on their own. These clusters do, perhaps, make a certain amount of sense. From the standardized scores shown in Table 8.1 it can be seen that Luxemburg is unusual because of the large numbers in mining. Spain is unusual because of the large numbers in

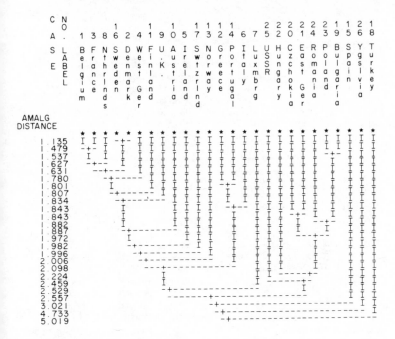

Figure 8.3 Dendrogram obtained from a nearest neighbour, hierarchic cluster analysis of data on employment in European countries.

construction. Yugoslavia is unusual because of the large numbers in agriculture and finance and low numbers in construction, social and personal services, and transport and communications. Turkey has extremely high numbers in agriculture and rather low numbers in most other areas.

An alternative analysis of the same data can be carried out using the BMDPKM program for a partitioning cluster analysis. This follows the iterative procedure described in Section 8.2 which starts with arbitrary cluster centres, allocates individuals to the closest centre, recalculates cluster centres, reallocates individuals, and so on. The number of clusters is at choice. For the data being considered, from two to six clusters were requested. With two clusters the program produced the following ones:

(1)	(2)
Belgium	Greece
Denmark	Spain
France	Turkey
West Germany	Bulgaria
Ireland	Czechoslovakia
Italy	East Germany
Luxemburg	Hungary
Netherlands	Poland
UK	Romania
Austria	USSR
Finland	Yugoslavia
Norway	Portugal
Sweden	
Switzerland	

For six clusters the choice was:

(1)	(2)	(3)	(4)	(5)	(6)
Luxemburg	East Germany	Turkey	Spain	Denmark	France
	Hungary	Yugoslavia	Bulgaria	Netherlands	West Germany
	Czechoslovakia		Poland	UK	Ireland
			Romania	Finland	Italy
			USSR	Norway	Austria
			Portugal	Sweden	Switzerland
			Greece	Belgium	

This is not the same as the six-cluster solution given by the dendrogram of Fig. 8.3, although there are some similarities. No doubt other algorithms for cluster analysis will give slightly different solutions.

The six clusters produced by the two methods of analysis can be compared with a plot of the countries against the first two principal components of the data (Fig. 5.2, p. 70). On this basis there is a fair but not perfect agreement: countries in the same cluster tend to have similar values for the first two principal components.

A comparison can also be made between the results of the cluster analyses and a plot of the countries against values for the first two canonical discriminant functions based on a division into EEC, non-EEC western European countries, and eastern European countries (Fig. 7.1, p. 95). Again there is a fair agreement, with countries in the same cluster tending to have similar values for the canonical discriminant functions.

Example 8.2 Relationships between canine species

As a second example, consider the data provided in Table 1.4 for mean mandible measurements of seven canine groups. As has been explained before, these data were originally collected as part of a study on the relationship between prehistoric dogs, whose remains have been uncovered in Thailand, and the other six living groups. This question has already been considered in terms of distances between the seven groups in Example 4.1. Table 4.1 (p. 45) shows mandible measurements standardized to have means of zero and standard deviations of one. Table 4.2 (p. 46) shows Euclidean distances between the groups based on these standardized measurements.

With only seven objects to be clustered it is simple to carry out a nearest-neighbour, hierarchic cluster analysis without using a computer. Thus it can be seen from Table 4.2 that the shortest distance between the seven groups in Example 4.1. Table 4.1 (p. 45) shows mandible measurements standardized to have means of zero and standard deviations of one. Table 4.2 (p. 45) shows Euclidean distances between the groups based on these standardized this distance level the cuon joins the prehistoric dog and modern dog in a cluster of three. The next largest distance is 1.63 between cuon and the modern dog. Since these two groups are already in the same cluster this has no effect. Continuing in this way produces the clusters at different distance levels that are shown in Table 8.2. The corresponding dendrogram is given in Figure 8.4.

Table 8.2 Clusters found at different distance levels for a hierarchic nearest-neighbour cluster analysis (MD = modern dog, GJ = golden jackal, CW = Chinese wolf, IW = Indian wolf, CU = cuon, DI = dingo and PD = prehistoric dog).

Distance		Number of clusters
0	MD, GJ, CW, IW, CU, DI, PD	7
0.72	(MD, PD), GJ, CW, IW, CU, DI	6
1.38	(MD, PD, CU), GJ, CW, IW, DI	5
1.63	(MD, PD, CU), GJ, CW, IW, DI	5
1.68	(MD, PD, CU, DI), GJ, CW, IW	4
1.80	(MD, PD, CU, DI), GJ, CW, IW	4
1.84	(MD, PD, CU, DI), GJ, CW, IW	4
2.07	(MD, PD, CU, DI, GJ), CW, IW	3
2.31	(MD, PD, CU, DI, GJ), (CW, IW)	2
2.37	(MD, PD, CU, DI, GJ, CW, IW)	1

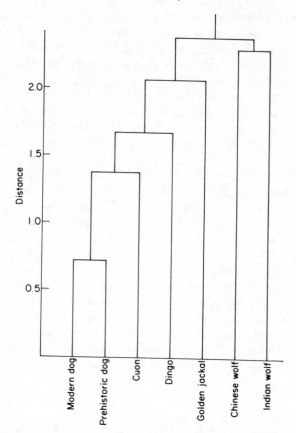

Figure 8.4 Dendrogram produced from the clusters shown in Table 8.2.

It appears that the prehistoric dog is closely related to the modern Thai dog, with both of these being somewhat related to the cuon and dingo and less closely related to the golden jackal. The Indian and Chinese wolves are closest to each other, but the difference between them is relatively large.

It seems fair to say that in this example the cluster analysis has produced a sensible description of the relationship between the different groups.

8.7 Further reading

There are a number of books available that provide more information about clustering methods and applications. Those by Aldenderfer and Blashfield (1984) and Gordon (1981) are at a level suitable for the novice. A more comprehensive account is given by Romesburg (1984).

References

Aldenderfer, M.S. and Blashfield, R.K. (1984) *Cluster Analysis.* Sage University Paper Series on Quantitative Applications in the Social Sciences, 07.001 Sage Publications, Beverly Hills.

Dixon, W.J. (1983) *BMDP Statistical Software.* University of California Press, Berkeley.

Gordon, A.D. (Ed.) (1981) *Classification.* Chapman and Hall, London.

Romesburg, H.C. (1984) *Cluster Analysis for Researchers.* Lifetime Learning Publications, Belmont, California.

CHAPTER NINE

Canonical correlation analysis

9.1 Generalizing a multiple regression analysis

In some sets of multivariate data the variables divide naturally into two groups. A canonical correlation analysis can then be used to investigate the relationships between the two groups. A case in point is the data that are provided in Table 1.3. Here 16 colonies of the butterfly *Euphydryas editha* in California and Oregon are considered. For each colony values are available for four environmental variables and six gene frequencies. An obvious question to be considered is what relationships, if any, exist between the gene frequencies and the environmental variables. One way to investigate this is through a canonical correlation analysis.

Another example was provided by Hotelling (1936) in one of the papers in which he described a canonical correlation analysis for the first time. This example involved the results of tests for reading speed (X_1), reading power (X_2), arithmetic speed (Y_1) and arithmetic power (Y_2) that were given to 140 seventh-grade schoolchildren. The specific question that was addressed was whether or not reading ability (as measured by X_1 and X_2) is related to arithmetic ability (as measured by Y_1 and Y_2). The approach that a canonical correlation analysis takes to answering this question is to search for a linear combination of X_1 and X_2, say

$$U = a_1 X_1 + a_2 X_2,$$

and a linear combination of Y_1 and Y_2, say

$$V = b_1 Y_1 + b_2 Y_2,$$

where these are chosen so that the correlation between U and V is as large as possible. This is somewhat similar to the idea in a principal component analysis except that here a correlation is maximized instead of a variance.

With X_1, X_2, Y_1 and Y_2 standardized to have unit variances, Hotelling found that the best choices for U and V are

$$U = -2.78X_1 + 2.27X_2 \quad \text{and} \quad V = -2.44Y_1 + 1.00Y_2,$$

where these have a correlation of 0.62. It can be seen that U measures the difference between reading power and speed and V measures the difference between arithmetic power and speed. Hence it appears that children with a large difference between X_1 and X_2 tended also to have a large difference between Y_1 and Y_2. It is this aspect of reading and arithmetic that shows most correlation.

It may be recalled that in a multiple regression analysis a single variable Y is related to two or more variables $X_1, X_2, \ldots, X_p$ to see how Y is related to the X's. From this point of view, canonical correlation analysis is a generalization of multiple regression in which several Y variables are simultaneously related to several X variables.

In practice more than one pair of canonical variates can be calculated from a set of data. If there are p variables $X_1, X_2, \ldots, X_p$ and q variables $Y_1, Y_2, \ldots, Y_q$ then there can be up to the minimum of p and q pairs of variates. That is to say, linear relationships

$$U_1 = a_{11}X_1 + a_{12}X_2 + \cdots + a_{1p}X_p$$
$$U_2 = a_{21}X_1 + a_{22}X_2 + \cdots + a_{2p}X_p$$
$$\vdots$$
$$U_r = a_{r1}X_1 + a_{r2}X_2 + \cdots + a_{rp}X_p$$

and

$$V_1 = b_{11}Y_1 + b_{12}Y_2 + \cdots + b_{1q}Y_q$$
$$V_2 = b_{21}Y_1 + b_{22}Y_2 + \cdots + b_{2q}Y_q$$
$$\vdots$$
$$V_r = b_{r1}Y_1 + b_{r2}Y_2 + \cdots + b_{rq}Y_q$$

can be established, where r is the smaller of p and q. These relationships are chosen so that the correlation between U_1 and V_1 is a maximum; the correlation between U_2 and V_2 is a maximum, subject to these variables being uncorrelated with U_1 and V_1; the correlation between U_3 and V_3 is a maximum, subject to these variables being uncorrelated with U_1, V_1, U_2 and V_2; and so on. Each of the pairs of canonical variables $(U_1, V_1), (U_2, V_2), \ldots, (U_r, V_r)$ then represents an independent 'dimension' in the relationship between the two sets of variables $(X_1, X_2, \ldots, X_p)$ and $(Y_1, Y_2, \ldots, Y_q)$. The first pair (U_1, V_1) have the highest possible correlation and are therefore the most important; the second pair (U_2, V_2) have the second highest correlation and are therefore the second most important; etc.

9.2 Procedure for a canonical correlation analysis

It is fairly easy to program the calculations for a canonical correlation analysis on a microcomputer, providing that suitable routines are available for matrix manipulations.

Assume that the $(p + q) \times (p + q)$ correlation matrix between the variables $X_1, X_2, \ldots, X_p, Y_1, Y_2, \ldots, Y_q$ takes the following form when it is calculated from the sample for which the variables are recorded:

$$
\begin{array}{c}
 \quad X_1 X_2 \ldots X_p \qquad Y_1 Y_2 \ldots Y_q \\
\begin{array}{c} X_1 \\ X_2 \\ \vdots \\ X_p \\[2pt] Y_1 \\ Y_2 \\ \vdots \\ Y_q \end{array}
\left[
\begin{array}{c|c}
\begin{array}{c} p \times p \text{ matrix} \\ \mathbf{A} \end{array} & \begin{array}{c} p \times q \text{ matrix} \\ \mathbf{C} \end{array} \\
\hline
\begin{array}{c} q \times p \text{ matrix} \\ \mathbf{C}' \end{array} & \begin{array}{c} q \times q \text{ matrix} \\ \mathbf{B} \end{array}
\end{array}
\right]
\end{array}
$$

From this matrix a $q \times q$ matrix $\mathbf{B}^{-1}\mathbf{C}'\mathbf{A}^{-1}\mathbf{C}$ can be calculated, and the eigenvalue problem

$$(\mathbf{B}^{-1}\mathbf{C}'\mathbf{A}^{-1}\mathbf{C} - \lambda\mathbf{I})\mathbf{b} = \mathbf{0} \qquad (9.1)$$

can be considered. It turns out that the eigenvalues $\lambda_1 > \lambda_2 > \cdots > \lambda_r$ are then the squares of the correlations between the canonical variates. The corresponding eigenvectors, $\mathbf{b}_1, \mathbf{b}_2, \ldots, \mathbf{b}_r$, give the coefficients of the Y variables for the canonical variates. The coefficients of U_i, the ith canonical variate for the X variables, are given by the elements of the vector

$$\mathbf{a}_i = \mathbf{A}^{-1}\mathbf{C}\mathbf{b}_i. \qquad (9.2)$$

In these calculations it is assumed that the original X and Y variables are in a standardized form with means of zero and standard deviations of unity. The coefficients of the canonical variates are for these standardized X and Y variables.

From equations (9.1) and (9.2) the ith pair of canonical variates are calculated as

$$U_i = \mathbf{a}_i'\mathbf{X} = (a_{i1}, a_{i2}, \ldots, a_{ip}) \begin{bmatrix} x_1 \\ x_2 \\ \vdots \\ x_p \end{bmatrix}$$

and

$$V_i = \mathbf{b}_i'\mathbf{Y} = (b_{i1}, b_{i2}, \ldots, b_{ip}) \begin{bmatrix} y_1 \\ y_2 \\ \vdots \\ y_q \end{bmatrix}$$

where $\mathbf{X}$ and $\mathbf{Y}$ are vectors of standardized data values. As they stand, U_i and V_i will have variances that depend upon the scaling adopted for the eigenvalue $\mathbf{b}_i$. However, it is a simple matter to calculate the standard deviation of U_i for the data and divide the a_{ij} values by this standard deviation. This produces a scaled canonical variate U_i with unit variance. Similarly, if the b_{ij} values are divided by the standard deviation of V_i then this produces a scaled V_i with unit variance.

This form of standardization of the canonical variates is not essential since the correlation between U_i and V_i is not affected by scaling. However, it may be useful when it comes to examining the numerical values of canonical variates for the individuals for which data is available.

9.3 Tests of significance

If there are r eigenvalues from equation (9.1) then there are r pairs of canonical variates. However, some of these may reflect correlations that are too small to be statistically significant. An approximate test proposed by Bartlett (1947) can be used to determine how many significant relationships exist.

To begin with, the test statistic

$$\phi_0^2 = -\left\{n - \tfrac{1}{2}(p + q + 1)\right\} \sum_{i=1}^{r} \log_e(1 - \lambda_i)$$

is calculated, where n is the number of cases for which data are available. This is compared with the percentage points of the chi-squared distribution with pq degrees of freedom. If ϕ_0^2 is significantly large then this establishes that there is at least one significant canonical correlation. If ϕ_0^2 is not significantly large then there is no evidence of any relationship between the X and Y variables.

Assuming that ϕ_0^2 is significant, the next step involves removing the effect of the first canonical correlation from the test statistic and considering

$$\phi_1^2 = -\left\{n - \tfrac{1}{2}(p + q + 1)\right\} \sum_{i=2}^{r} \log_e(1 - \lambda_i),$$

with $(p - 1)(q - 1)$ degrees of freedom. If this is significantly large in comparison with the chi-squared percentage points then there are at least two significant canonical correlations. If ϕ_1^2 is not significantly large then the first canonical correlation can be considered to account for all of the relationships between the X and Y variables.

If ϕ_0^2 and ϕ_1^2 are both significant then the effect of the first two canonical correlations can be removed from the test statistic to see if any of the remaining correlations are significant. This process can continue until it is found that the remaining correlations are no longer significant and hence can be neglected. The test statistic for the remaining correlations after the first j have been removed is

$$\phi_j^2 = -\left\{n - \tfrac{1}{2}(p + q + 1)\right\} \sum_{i=j+1}^{r} \log_e(1 - \lambda_i), \tag{9.3}$$

with $(p - j)(q - j)$ degrees of freedom.

Some texts on multivariate analysis provide details of modifications to these test statistics to improve the approximations involved for small sample sizes n. However, the formulae stated here are relatively simple and are ones that are often used.

9.4 Interpreting canonical variates

Having decided which canonical variates, if any, are significant for a set of data, the next problem is the interpretation of these variates: what are they measuring? At first sight it may seem that this is a relatively easy question to answer. If

$$U_i = a_{i1}X_1 + a_{i2}X_2 + \cdots + a_{ip}X_p$$

and

$$V_i = b_{i1}Y_1 + b_{i2}Y_2 + \cdots + b_{iq}Y_q$$

then it seems that U_i can be described in terms of the X variables with large coefficients a_{ij} and V_i can be described in terms of the Y variables with large coefficients b_{ij}. 'Large' here means, of course, large positive or large negative.

Unfortunately, correlations between the X and Y variables can upset this interpretation process. For example, it can happen that a_{i1} is positive and yet the simple correlation between U_i and X_1 is actually negative. This apparent contradiction can come about because X_1 is highly correlated with one or more of the other X variables and part of the effect of X_1 is being accounted for by the coefficients of these other X variables. In fact, if one of the X variables is almost a linear combination of the other X variables then there will be an infinite variety of linear combinations of the X variables, some of them with very different a_{ij} values, that give virtually the same U_i values. The same can be said about linear combinations of Y variables.

The interpretation problems that arise with high correlated X variables or Y variables should be familiar to users of multiple regression analysis. Exactly the same problems arise with the estimation of regression coefficients.

Actually, a fair comment seems to be that if the X or Y variables are highly correlated then there can be no way of disentangling their contributions to canonical variates. However, no doubt people will continue to try to make interpretations under these circumstances.

Some authors have suggested that it is better to describe canonical variates by looking at their correlations with the X and Y variables rather than the coefficients a_{ij} and b_{ij}. For example, if U_i is highly positively correlated with X_1 then U_i can be considered to reflect X_1 to a large extent. Similarly, if V_i is highly negatively correlated with Y_1 then V_i can be considered to reflect the opposite of Y_1 to a large extent. This approach does at least have the merit of bringing out all of the variables to which the canonical variates seem to be related.

Example 9.1 Environmental and genetic correlations · for colonies of Euphydryas editha

The data in Table 1.3 can be used to illustrate the procedure for a canonical correlation analysis. Here there are 16 colonies of the butterfly *Euphydryas editha* in California and Oregon. These vary with respect to four environmental variables (altitude, annual precipitation, annual maximum temperature, and annual minimum temperature) and six genetic variables (percentages of six phosphoglucose-isomerase genes as determined by electrophoresis). Any significant relationships between the environmental and genetic variables are interesting because they may indicate the adaption of *E. editha* to the environment.

For reasons to be explained in Section 9.5 on computational methods, there may be some advantage in labelling variables so that the number of X variables (p) is greater than or equal to the number of Y variables (q) in a canonical correlation analysis. In the present case this suggests that the X variables should be the gene frequencies. Since these frequencies add to exactly 100% there is no point in including them all in the analysis. The last frequency, which is the percentage of genes with a mobility of 1.30, will therefore be omitted on the grounds that it can be written as 100% minus the sum of the other percentages. Thus the X variables to be considered are $X_1 =$ percentage of 0.40 mobility genes, $X_2 =$ percentage of 0.60 mobility genes, $X_3 =$ percentage of 0.80 mobility genes, $X_4 =$ percentage of 1.00 mobility genes, and $X_5 =$ percentage of 1.16 mobility genes. The Y variables are $Y_1 =$ altitude, $Y_2 =$ annual percipitation, $Y_3 =$ annual maximum temperature and $Y_4 =$ annual minimum temperature. It should be noted that there is no need to think of the X variables as 'causing' the Y variables, or vice versa. From this point of view the labelling of variables as X's or Y's is arbitrary.

The standardized variables are shown in Table 9.1. These produce the correlation matrix shown in Table 9.2, which is partitioned into the $\mathbf{A}$, $\mathbf{B}$, $\mathbf{C}$ and $\mathbf{C}'$ matrices as described in Section 9.2

The eigenvalues obtained from equation (9.1) are 0.7731, 0.5570, 0.1694 and 0.0472. Taking square roots gives the canonical correlations that these provide, 0.879, 0.746, 0.412 and 0.217. The corresponding canonical variates are obtained from equations (9.1) and (9.2). After standardizing to have unit variances these become:

$$U_1 = -0.675X_1 + 0.909X_2 + 0.376X_3 + 1.442X_4 + 0.269X_5,$$
$$V_1 = -0.114Y_1 + 0.619Y_2 - 0.693Y_3 + 0.048Y_4,$$

$$U_2 = -1.087X_1 + 3.034X_2 + 2.216X_3 + 3.439X_4 + 2.928X_5$$
$$V_2 = -0.777Y_1 + 0.980Y_2 - 0.562Y_3 + 0.928Y_4,$$

$$U_3 = 1.530X_1 + 2.049X_2 + 2.231X_3 + 4.916X_4 + 3.611X_5,$$
$$V_3 = -3.654Y_1 - 0.601Y_2 - 0.565Y_3 - 3.623Y_4,$$

$$U_4 = 0.284X_1 - 2.331X_2 - 0.867X_3 - 1.907X_4 - 1.133X_5,$$
$$V_4 = 1.594Y_1 + 0.860Y_2 + 1.599Y_3 + 0.742Y_4.$$

Although the first two canonical correlations are quite large, they are not significantly large by Bartlett's test. The test statistic for the significance of the canonical variates as a whole is found from equation (9.3) to be $\phi_0^2 = 27.85$ with 20 degrees of freedom. Compared to the chi-squared distribution this is not at all significantly large. The test statistic for canonical variates 2, 3 and 4 is found to be $\phi_1^2 = 11.53$ with 12 degrees of freedom. The test statistics for canonical variates 3 and 4 ($\phi_2^2 = 2.57$, with 6 degrees of freedom) and for canonical variate 4 alone ($\phi_3^2 = 0.52$ with 2 degrees of freedom) are even less significant. It appears, therefore, that there is no real evidence of any relationships between the genetic and environmental variables.

It may seem strange that there is no significance in the results although the first canonical correlation of 0.879 is quite high. The explanation is the rather small sample size of 16.

Laying aside the lack of significance, it is interesting to see what

Table 9.1 Standardized values for X_1 to X_5 and Y_1 to Y_4 for 16 colonies of *Euphydryas editha*. The unstandardized variables X_1 to X_5 are the percentages of Pgi genes with different electrophoretic mobilities. They are the first five of the six frequencies given in Table 1.3. The unstandardized variables Y_1 to Y_4 are altitude, annual precipitation, annual maximum temperature, and annual minimum temperature, respectively, again as given in Table 1.3.

Colony	Genetic variables					Environmental variables			
	X_1	X_2	X_3	X_4	X_5	Y_1	Y_2	Y_3	Y_4
1	-0.42	-0.56	0.32	0.29	-0.02	-0.56	1.07	0.12	-0.35
2	-0.42	1.18	0.13	-0.66	-0.38	-0.45	-0.58	-0.82	1.02
3	-0.42	-0.16	0.88	-0.26	-0.02	-0.53	0.00	0.12	0.47
4	-0.42	-0.43	0.04	-0.21	0.91	-0.54	0.00	0.12	0.47
5	-0.42	-0.83	-0.98	-0.06	1.65	-0.54	0.00	0.12	0.47
6	-0.42	-0.69	0.04	-0.36	1.37	-0.60	-0.93	0.27	0.65
7	-0.42	-0.96	-0.33	-0.06	0.91	-0.41	-0.50	0.27	0.65
8	1.99	1.84	1.99	-1.31	-1.22	-0.50	-1.29	0.59	0.56
9	2.96	2.51	1.25	-1.16	-1.59	-0.52	-1.29	0.59	0.56
10	-0.42	-0.83	-1.17	1.44	-0.48	-0.21	-0.65	0.27	0.19
11	-0.18	-0.43	1.43	-0.91	0.45	-0.12	-0.43	0.59	0.56
12	-0.42	-0.03	-0.42	0.74	-0.39	-0.04	2.14	0.43	-0.26
13	-0.42	0.24	-0.33	-0.21	0.35	0.14	0.42	0.74	-0.44
14	0.30	-0.03	-0.14	-0.96	0.91	-0.04	-0.50	1.21	-0.08
15	-0.42	-0.29	-1.07	1.64	-1.22	2.00	1.00	-2.08	-1.45
16	-0.42	-0.56	-1.63	2.03	-1.22	2.92	1.57	-2.55	-3.00

Table 9.2 Correlation matrix for the variables X_1 to X_5 and Y_1 to Y_4 for the *Euphydryas editha* data, partitioned into **A**, **B**, **C** and **C'** matrices.

		A					C	
1.000	0.855	0.618	-0.532	-0.506	-0.203	-0.530	0.295	0.221
0.855	1.000	0.615	-0.548	-0.597	-0.190	-0.410	0.173	0.246
0.618	0.615	1.000	-0.824	-0.127	-0.573	-0.550	-0.536	0.593
-0.532	-0.548	-0.824	1.000	-0.264	0.727	0.699	-0.717	-0.759
-0.506	-0.597	-0.127	-0.264	1.000	-0.458	-0.138	0.438	0.412
		C'					B	
-0.203	-0.190	-0.573	0.727	-0.458	1.000	0.568	-0.828	-0.936
-0.530	-0.410	-0.550	0.699	-0.138	0.568	1.000	-0.479	-0.705
0.295	0.173	0.536	-0.717	0.438	-0.828	-0.479	1.000	0.719
0.221	0.246	0.593	-0.759	0.412	-0.936	-0.705	0.719	1.000

interpretation can be given to the first pair of canonical variates (U_1, V_1). From the equation for U_1 it appears that this is a contrast between X_1 and the other X variables. It represents a lack of genes with mobility 0.40. On the other hand, V_1 has a large positive coefficient for Y_2 (precipitation) and a large negative coefficient for Y_3 (maximum temperature). It would seem that the 0.40 mobility gene is lacking in colonies with high precipitation and low maximum temperatures.

The correlations between U_1 and the five X variables are as follows: U_1 and X_1, -0.57; U_1 and X_2, -0.39; U_1 and X_3, -0.70; U_1 and X_4, 0.92; U_1 and X_5, -0.36. Thus U_1 is highly positively correlated with X_4 (the percentage of mobility 1.00 genes) and negatively correlated with the other X variables. This suggests that U_1 is best interpreted as indicating a high frequency of mobility 1.00 genes. This is a somewhat different interpretation from the one given by a consideration of the coefficient of U_1 for the X variables. On the whole, the interpretation based on correlations seems best. However, as mentioned in the previous section, there are real problems about interpreting canonical variates when the variables that they are constructed from have high correlations. Table 9.2 shows that this is indeed the case with the present example.

The correlation between V_1 and the four Y variables are as follows: V_1 and Y_1, 0.77; V_1 and Y_2, 0.85; V_1 and Y_3, -0.86; V_1 and Y_4, -0.78. Thus V_1 seems to be associated with high altitude and precipitation and low temperatures.

Taken together, the interpretation of U_1 and V_1 based on correlations suggests that the percentage of mobility 1.00 genes is high for colonies with high altitudes and precipitation and low temperatures. This does indeed show up in the data to some extent. For example, colony 16 has the highest value of X_4, high values for Y_1 and Y_2, and low values for Y_3 and Y_4 (Table 9.1).

9.5 Computational methods

A canonical correlation analysis involves the inversion of a matrix and the solution of the eigenvalue problem of equation (9.1). This equation can be rewritten as

$$(\mathbf{C}'\mathbf{A}^{-1}\mathbf{C} - \lambda\mathbf{B})\mathbf{b} = \mathbf{0}.$$

Here $C'A^{-1}C$ and B are symmetric and hence the eigenvalues and eigenvectors can be found using Algorithm 15 of Nash (1979). This approach requires that the number of positive eigenvalues is equal to q, the number of rows and columns in the matrices $C'A^{-1}C$ and B. This will not be the case if there are more Y variables than X variables in a canonical correlation analysis. Hence the variables should be labelled so that the number of Y variables is less than or equal to the number of X variables. This is no real problem since there is no implication in a canonical correlation analysis that one of the sets of variables is dependent on the other set, although this may sometimes be the case.

The inversion of the matrix A can be done using either Algorithm 5 or Algorithm 9 of Nash (1979).

9.6 Further reading

Giffins (1985) has written a book on canonical correlation intended mainly for ecologists. About half of the work is devoted to theory and methods, and the remainder to a number of specific examples in the area of plant ecology. Less detailed introductions are provided by Clarke (1975) and Levine (1977).

References

Bartlett, M.S. (1947) The general canonical correlation distribution. *Annals of Mathematical Statistics* **18**, 1–17.

Clarke, D. (1975) *Understanding Canonical Correlation Analysis*. Concepts and Techniques in Modern Geography 3, Geo. Abstracts, Norwich, UK.

Giffins, R. (1985) *Canonical Analysis: A Review with Applications in Ecology*. Biomathematics 12, Springer-Verlag, Berlin.

Hotelling, H. (1936) Relations between two sets of variables. *Biometrika* **28**, 321–77.

Levine, M.S. (1977) *Canonical Analysis and Factor Comparisons*. Sage University Papers on Quantitative Applications in the Social Sciences 07-006. Sage Publications, Beverly Hills.

Nash, J.C. (1979) *Compact Numerical Methods for Computers*. Adam Hilger, Bristol.

CHAPTER TEN

Multidimensional scaling

10.1 Constructing a 'map' from a distance matrix

Multidimensional scaling is a technique that is designed to construct a 'map' showing the relationships between a number of objects, given only a table of distances between them. The 'map' can be in one dimension (if the objects fall on a line), in two dimensions (if the objects lie on a plane), in three dimensions (if the objects can be represented by points in space), or in a higher number of dimensions (in which case an immediate geometrical representation is not possible).

The fact that it may be possible to construct a map from a table of distances can be seen by considering the example of four objects A, B, C and D shown in Fig. 10.1. Here the distances apart are given by the array:

	A	B	C	D
A	0	6.0	6.0	2.5
B	6.0	0	9.5	7.8
C	6.0	9.5	0	3.5
D	2.5	7.8	3.5	0

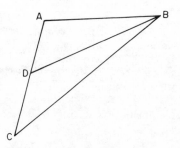

Figure 10.1 A map of the relationship between four objects.

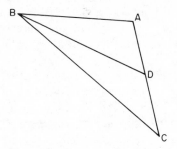

Figure 10.2 A mirror image of the map shown in Fig. 10.1 for which the distances between objects are the same.

For example, the distance from A to B, which is the same as the distance from B to A, is 6.0. The distance of objects to themselves is, of course, 0. It seems plausible that the map can be reconstructed from the array of distances. A moments reflection may indicate, however, that a mirror image of the map as shown in Fig. 10.2 will have the same array of distances between objects. Consequently, it seems clear that a recovery of the original map will be subject to a possible reversal of this type.

It is also apparent that if more than three objects are involved then they may not lie on a plane. In that case their distance matrix will implicitly contain this information. For example, the distance array:

	A	B	C	D
A	0	1	$\sqrt{2}$	$\sqrt{2}$
B	1	0	1	1
C	$\sqrt{2}$	1	0	$\sqrt{2}$
D	$\sqrt{2}$	1	$\sqrt{2}$	0

is such that three dimensions are required to show the spatial relationships between the four objects. Unfortunately, with real data it is not usually known how many dimensions are needed for a representation. Hence a range of dimensions has to be tried.

The usefulness of multidimensional scaling comes from the fact that situations often arise where the relationship between objects is not known, but a distance matrix can be estimated. This is particularly the case in psychology where subjects can say how similar or different individual pairs of objects are, but they cannot draw an overall picture of the relationships between the objects. Multidimensional scaling can then provide a picture. The main applications to date have been in psychology and sociology.

At the present time there are a wide variety of data analysis techniques that go under the general heading of multidimensional scaling. Here only the simplest of these will be considered, these being the classical methods proposed by Torgerson (1952) and Kruskal (1964).

10.2 Procedure for multidimensional scaling

A classical multidimensional scaling starts with a matrix of distances between n objects which has δ_{ij}, the distance from object i to object j, in the ith row and jth column. The number of dimensions, t, for the mapping of objects is fixed for a particular solution. Different computer programs use different methods for carrying out analysis but generally something like the following steps are involved:

1. A starting configuration is set up for the n objects in t dimensions, i.e. coordinates $(x_1, x_2, \ldots, x_t)$ are assumed for each object in a t-dimensional space.
2. The Euclidean distances between the individuals are calculated for the configuration. Let d_{ij} be the distance between individual i and individual j.
3. A regression of d_{ij} on δ_{ij} is made where, as mentioned above, δ_{ij} is the distance between individual i and individual j according to the input data. The regression can be linear, polynomial or monotonic. For example, a linear regression assumes that

$$d_{ij} = a + b\delta_{ij} + e,$$

where e is an 'error' term and a and b are constants. A monotonic regression assumes simply that if δ_{ij} increases then d_{ij} increases or remains constant but no exact relationship between δ_{ij} and d_{ij} is fitted. The distances obtained from the regression equation ($\hat{d}_{ij} = a + b\delta_{ij}$, assuming a linear regression) are called 'disparities'. That is to say, the disparities $\hat{d}_{ij}$ are the data distances δ_{ij} scaled to match the configuration distance d_{ij} as closely as possible.
4. The goodness of fit between the configuration distances and the disparities is measured by a suitable statistic. One possibility is

called 'stress formula 1', which is

$$\text{STRESS } 1 = \left\{ \sum (d_{ij} - \hat{d}_{ij})^2 / \sum \hat{d}_{ij}^2 \right\}^{1/2} \tag{10.1}$$

The description 'stress' is used since the statistic is a measure of the extent to which the spatial configuration of points has to be stressed in order to obtain the data distances δ_{ij}.

5. The coordinates $(x_1, x_2, \ldots, x_t)$ of each object are changed slightly in such a way that the stress is reduced.

Steps 2 to 5 are repeated until it seems that the stress cannot be further reduced. The outcome of the analysis is then the coordinates of the n individuals in t dimensions. These coordinates can be used to draw a 'map' which shows how the individuals are related.

It is desirable that a good solution is found in three or fewer dimensions, since a graphical representation of the n objects is then straightforward. Obviously this is not always possible.

Example 10.1 Road distances between New Zealand towns

As an example of what can be achieved by multidimensional scaling, a 'map' of the south island of New Zealand, has been constructed from a table of the road distances between the 13 towns shown in Fig. 10.3.

If road distances were proportional to geographic distances it would be possible to recover the true map exactly by a two-dimensional analysis. However, due to the absence of direct road links between many towns, road distances are in some cases far greater than geographic distances. Consequently, all that can be hoped for is a rather approximate recovery.

The computer program ALSCAL-4 (Young and Lewyckyj, 1979) was used for the analysis. This is one of several widely available programs for multidimensional scaling. At step 3 of the procedure described above a monotonic regression relationship was assumed between map distances d_{ij} and the distances δ_{ij} given in Table 10.1. That is, it was assumed that an increase in δ_{ij} implies that d_{ij} either increases or remains constant. The analysis was therefore what is sometimes called classical non-metric multidimensional scaling.

The program produced a two-dimensional solution for the data in four iterations of steps 2 to 5 of the algorithm described above. The final stress value was 0.052 as calculated using equation (10.1).

The output from the program includes the coordinates of the 13

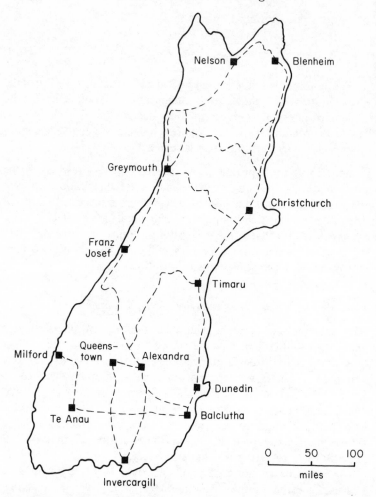

Figure 10.3 The south island of New Zealand. Main roads are indicated by broken lines. The 13 towns used for example 10.1 are indicated.

towns on the 'map' produced in the analysis. These are shown in Table 10.2. A plot of the towns using these coordinates is shown in Fig. 10.4. A comparison of this figure with Fig. 10.3 indicates that the multidimensional scaling has been quite successful in recovering the map of South Island. On the whole the towns are shown with the correct relationships to each other. An exception is Milford. Because

Table 10.1 Main road distances (miles) between 13 towns in the south island of New Zealand. The positions of the towns and road links are shown in Fig. 10.3.

	Alexandra	Balclutha	Blenheim	Christ-church	Dunedin	Franz Josef	Grey-mouth	Inver-cargill	Milford	Nelson	Queens-town	Te Anau	Timaru
Alexandra	–												
Balclutha	100	–											
Blenheim	485	478	–										
Christchurch	284	276	201	–									
Dunedin	126	50	427	226	–								
Franz Josef	233	493	327	247	354	–							
Greymouth	347	402	214	158	352	114	–						
Invercargill	138	89	567	365	139	380	493	–					
Milford	248	213	691	489	263	416	555	174	–				
Nelson	563	537	73	267	493	300	187	632	756	–			
Queenstown	56	156	494	305	192	228	341	118	178	572	–		
Te Anau	173	138	615	414	188	366	480	99	75	681	117	–	
Timaru	197	177	300	99	127	313	225	266	377	366	230	315	–

Table 10.2 Coordinates produced by multi-dimensional scaling applied to the distances between 13 towns shown in Table 10.1. These are the coordinates that the towns are plotted against in Fig. 10.4.

Town	Dimension 1	Dimension 2
Alexandra	0.72	−0.32
Balclutha	0.84	0.78
Blenheim	−1.99	0.43
Christchurch	−0.92	0.34
Dunedin	0.52	0.46
Franz Josef	−0.69	−1.23
Greymouth	−1.32	−0.57
Invercargill	1.28	0.39
Milford	1.83	−0.33
Nelson	−2.33	0.07
Queenstown	0.81	−0.49
Te Anau	1.47	−0.26
Timaru	−0.19	0.64

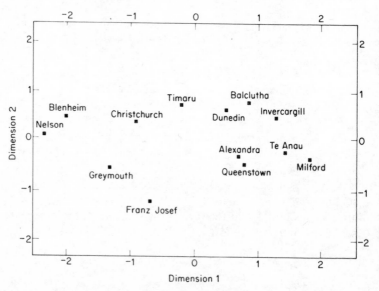

Figure 10.4 The map produced by a multidimensional scaling analysis of the data in Table 10.1.

this can only be reached by road through Te Anau, the 'map' produced by multidimensional scaling has made Milford closest to Te Anau. In fact, Milford is geographically closer to Queenstown than it is to Te Anau.

All that is important with the configuration produced by multidimensional scaling is the relative positions of the objects being considered. This is unchanged by a rotation or a reflection. It is also unchanged by a magnification or contraction of all the scales. That is, the size of the configuration is not important. For this reason ALSCAL-4 always scales the configuration so that the average coordinate is zero in all dimensions and the sum of the squared coordinates is equal to the number of objects multiplied by the number of dimensions. Thus in Table 10.2 the sum of the coordinates is zero for each of the two dimensions and the total of the coordinates squared is 26.

Example 10.2 Voting behaviour of Congressmen

For a second example of the value of multidimensional scaling, consider the distance matrix shown in Table 10.3. Here the 'distances' are between 15 New Jersey Congressmen in the United States House of Representatives. They are simply a count of the number of voting disagreements on 19 bills concerned with environmental matters. For example, Congressmen Hunt and Sandman disagreed 8 out of the 19 times. Sandman and Howard disagreed 17 out of the 19 times, etc. An agreement was considered to occur if two Congressmen both voted yes, both voted no, or both failed to vote. The table of distances was constructed from original data given by Romesburg (1984), p. 155).

Two analyses were carried out using the ALSCAL-4 program. The first was a classical metric multidimensional scaling which assumes that the distances of Table 10.3 are measured on a ratio scale. That is to say, it is assumed that doubling a distance value is equivalent to assuming that the configuration distance between two objects is doubled. This means that the regression at step 3 of the procedure described above is of the form

$$d_{ij} = b\delta_{ij} + e,$$

where e is an error and b is some constant. The stress values obtained for 4-, 3- and 2-dimensional solutions were found on this basis to be 0.080, 0.121 and 0.194, respectively.

Table 10.3 The 'distances' between 15 Congressmen from New Jersey in the United States House of Representatives. The numbers in the table show the number of times that the Congressmen voted differently on 19 environmental bills. Party allegiancies are indicated (R = Republican, D = Democrat).

	1	2	3	4	5	6	7	8	9	10	11	12	13	14	15
1 Hunt (R)	—														
2 Sandman (R)	8	—													
3 Howard (D)	15	17	—												
4 Thompson (D)	15	12	9	—											
5 Frelinghuysen (R)	10	13	16	14	—										
6 Forsythe (R)	9	13	12	12	8	—									
7 Widnall (R)	7	12	15	13	9	7	—								
8 Roe (D)	15	16	5	10	13	12	17	—							
9 Helstoski (D)	16	17	5	8	14	11	16	4	—						
10 Rodino (D)	14	15	6	8	12	10	15	5	3	—					
11 Minish (D)	15	16	5	8	12	9	14	5	2	1	—				
12 Rinaldo (R)	16	17	4	6	12	10	15	3	1	2	1	—			
13 Maraziti (R)	7	13	11	15	10	6	10	12	13	11	12	12	—		
14 Daniels (D)	11	12	10	10	11	6	11	7	7	4	5	5	9	—	
15 Patten (D)	13	16	7	7	11	10	13	6	5	6	5	4	13	9	—

A second analysis was carried out by classical non-metric scaling so that the regression of d_{ij} on δ_{ij} was assumed to be monotonic only. In this case the stress values for 4-, 3- and 2-dimensional solutions were found to be 0.065, 0.089 and 0.134, respectively. The distinctly lower stress values for non-metric scaling suggest that this is preferable to metric scaling for these data. The three-dimensional non-metric solution has only slightly more stress than the four-dimensional solution. This three-dimensional solution is therefore the one that will be considered in more detail.

Table 10.4 shows the coordinates of the Congressmen for the three-dimensional solution. A plot for the first two dimensions is shown in Fig. 10.5. The value for the third dimension is shown for each plotted point in Fig. 10.5, where this dimension indicates how far a three-dimensional plot would make the point above or below the two-dimensional plane. For example, Daniels should be plotted 0.52 units above the plane and Rinaldo should be plotted 0.27 units below the plane.

From Fig. 10.5 it is clear that dimension 1 is largely reflecting party

Table 10.4 Coordinates of the 15 Congressmen obtained from a three-dimensional non-metric multidimensional scaling of the distance matrix given in Table 10.3.

| Congressmen | Dimension | | |
	1	2	3
Hunt	2.25	0.15	0.53
Sandman	1.74	2.06	0.64
Howard	− 1.37	− 0.01	0.84
Thompson	− 0.85	1.42	− 0.45
Frelinghuysen	1.47	− 0.83	− 1.23
Forsythe	0.81	− 0.93	− 0.43
Widnall	2.25	− 0.28	− 0.46
Roe	− 1.40	− 0.01	0.60
Helstoski	− 1.50	0.22	− 0.18
Rodino	− 1.09	− 0.19	0.10
Minish	− 1.13	− 0.21	− 0.24
Rinaldo	− 1.27	− 0.18	− 0.27
Maraziti	1.20	− 1.20	0.97
Daniels	− 0.12	− 0.16	0.52
Patten	− 0.99	0.14	− 0.94

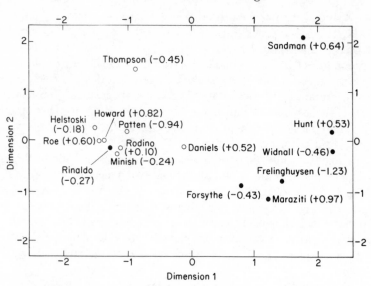

Figure 10.5 Plot of Congressmen against the first two dimensions of the configuration produced by a three-dimensional classical non-metric multidimensional scaling of the data in Table 10.3. Open circles indicate Democrats, closed circles indicate Republicans. The coordinate for dimension 3 is indicated in parenthesis for each point.

differences. The Democrats fall on the left-hand side of the figure and the Republicans, other than Rinaldo, on the right-hand side.

To interpret dimension 2 it is necessary to consider what it is about the voting of Sandman and Thompson, who have the highest two scores, that contrasts with Maraziti and Forsythe, who have the two lowest scores. This points to the number of abstentions from voting. Sandman abstained from nine votes and Thompson abstained from six votes. Individuals with low scores on dimension 2 voted all or most of the time.

Dimension 3 appears to have no simple or obvious interpretation. It must reflect certain aspects of differences in voting patterns. However, these will not be considered for the present example. It suffices to say that the analysis has produced a representation of the Congressmen in three dimensions that indicates how they relate with regard to voting on environmental issues.

Figure 10.6 Plot of configuration distances against disparities (scaled data distances). One point is indicated by an X, multiple points by a digit.

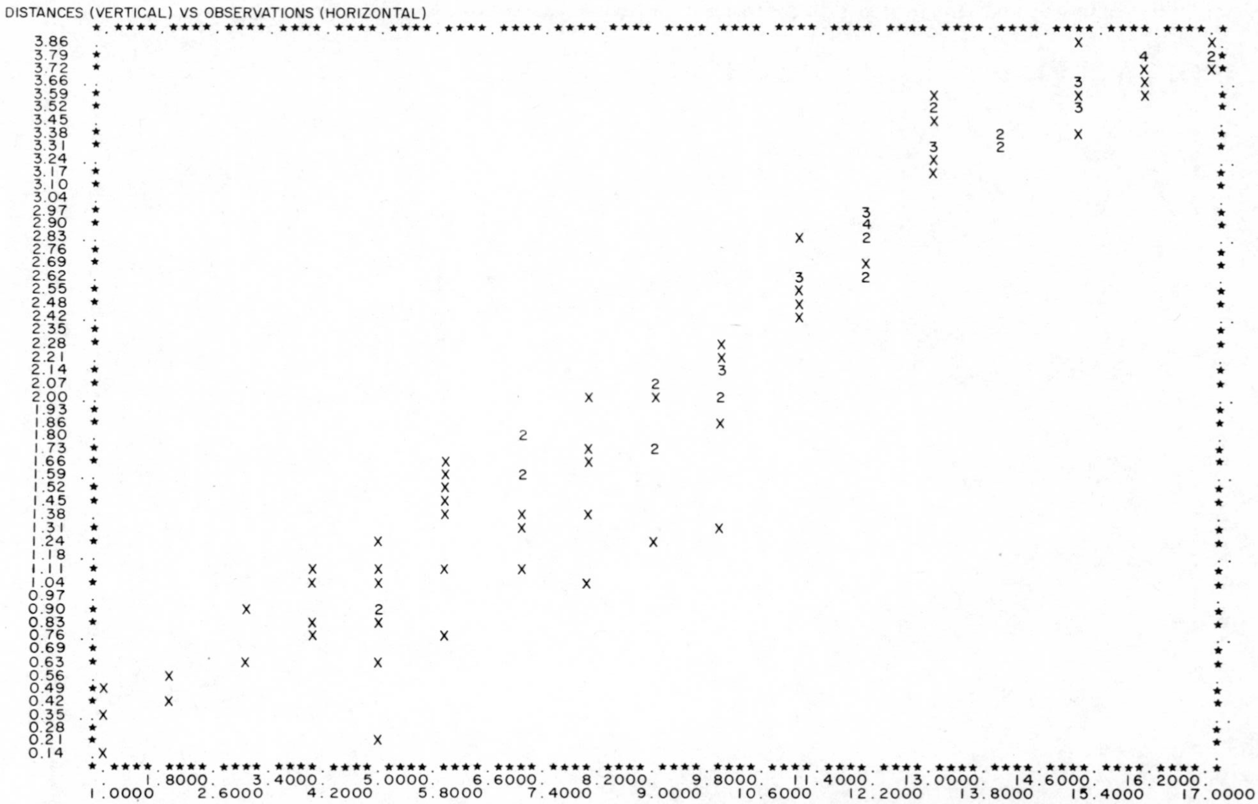

Figure 10.7 Plot of configuration distances against the corresponding distances in the data.

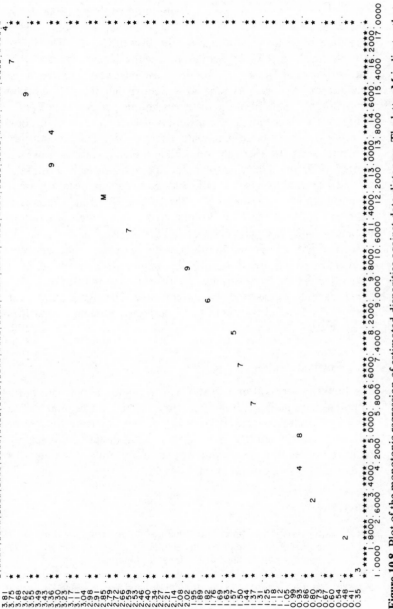

Figure 10.8 Plot of the monotonic regression of estimated disparities against data distances. The letter M indicates that more than nine points fall in the same position. In fact there are twelve distances of 12 in Table 10.3 that all yield estimated discrepancies of about 2.85.

Three graphs output by ALSCAL-4 are helpful in assessing the accuracy of the solution that has been obtained. Figure 10.6 shows the first of these, which is a plot of the distances between points on the derived configuration, d_{ij}, against the disparities, $\hat{d}_{ij}$. The figure indicates the lack of fit of the solution since the disparities are the 105 data distances of Table 10.3 after they have been scaled to match the configuration as closely as possible. A plot like that of Fig. 10.6 would be a straight line if all of the distances and disparities were equal.

Figure 10.7 is a plot of the distances between the configuration points (d_{ij}) against the original data distances (δ_{ij}). This relationship does not have to be a straight line with non-metric scaling. However, scatter about an underlying trend line does indicate lack of fit of the model. For example in Table 10.3 there are eight distances of 5. Figure 10.7 shows that these correspond to configuration distances of between about 0.2 and 1.25. With an error-free solution, equal data distances will give equal configuration distances.

Finally, Fig. 10.8 shows the monotonic regression that has been estimated between the disparities $\hat{d}_{ij}$ and the data distances δ_{ij} of Table 10.3. As the data distances increase, all that is required is that the disparities increase or remain constant. The figure shows, for example, that data distances of 4 or 5 both give estimated disparities of about 0.93.

10.3 Further reading

The book by Kruskal and Wish (1978) provides a short but fairly comprehensive guide to multidimensional scaling for beginners. The book by Schiffman *et al.* (1981) is longer but describes the use of several available computer programs. Multidimensional scaling requires the use of a good computer program. Both of these books have lists of sources of programs.

References

Kruskal, J.B. (1964) Multidimensional scaling by optimizing goodness of fit to a nonmetric hypothesis *Psychometrika* **29**, 1–27.
    Nonmetric multidimensional scaling: a numerical method. *Psychometrika*, **29**, 115–29.
Kruskal, J.B. and Wish, M. (1978) *Multidimensional Scaling.* Sage University Paper Series on Quantitative Applications in the Social Sciences, 07-011. Sage Publications, Beverly Hills.

Romesburg, H.C. (1984) *Cluster Analysis for Researchers*. Lifetime Learning Publications, Belmont, California.

Schiffman, S.S., Reynolds, M.L. and Young, F.W. (1981) *Introduction to Multidimensional Scaling*. Academic Press, Orlando, Florida.

Torgerson, W.S. (1952) Multidimensional scaling. I. Theory and method. *Psychometrika* **17**, 401–19.

Young, F.W. and Lewyckyj, R. (1979) *ALSCAL-4 User's Guide*. Psychometric Laboratory, University of North Carolina, Chapel Hill.

CHAPTER ELEVEN

Epilogue

11.1 The next step

In writing this book my aims have purposely been rather limited. These aims will have been achieved if someone who has read the previous chapters carefully has a fair idea of what can and what cannot be achieved by the multivariate statistical methods that are most widely used. My hope is that the book will help many people take the first step in 'a journey of a thousand miles'.

For those who have taken this first step, the way to proceed further is to gain experience of multivariate methods by analysing different sets of data and seeing what results are obtained. It will be very helpful, if not essential, to get access to one of the larger statistical packages and investigate the different options that are available. Like other areas of applied statistics, competence in multivariate analysis requires practice. To this end, the Appendix contains some sets of data with suggestions about how to examine them.

11.2 Some general reminders

In developing expertise and familiarity with multivariate analyses there are a few general points that are worth keeping in mind. Actually, these points are just as relevant to univariate analyses. However, they are still worth emphasizing in the multivariate context.

First, it should be remembered that there are often alternative ways of approaching the analysis of a particular set of data, none of which is necessarily the 'best'. Indeed, several types of analysis may well be carried out to investigate different aspects of the same data. For example, the body measurements of female sparrows given in Table 1.1 can be analysed by principal component analysis or factor analysis to investigate the dimensions behind body size variation, by discriminant analysis to contrast survivors and non-survivors, by cluster analysis or multidimensional scaling to see how the birds group together, and so on.

Second, use common sense. Before embarking on an analysis consider whether it can possibly answer the questions of interest. How many statistical analyses are carried out because the data are of the right form, irrespective of what light the analyses can throw on a question? At some time or another most users of statistics find themselves sitting in front of a large pile of computer output with the realization that it tells them nothing that they really want to know.

Third, multivariate analysis does not always work in terms of producing a 'neat' answer. There is an obvious bias in statistical textbooks and articles towards examples where results are straightforward and conclusions are clear. In real life this does not happen quite so often. Do not be surprised if multivariate analyses fail to give satisfactory results on the data that you are really interested in! It may well be that the data have message to give, but the message cannot be read using the somewhat simple models that standard analyses are based on. For example, it may be that variation in a multivariate set of data can be completely described by two or three underlying factors. However, these may not show up in a principal component analysis or a factor analysis because the relationship between the observed variables and the factors is not a simple linear one.

Finally, there is always the possibility that an analysis is dominated by one or two rather extreme observations. These 'outliers' can sometimes be found by simply scanning the data by eye, or by considering frequency tables for the distributions of individual variables. In some cases a more sophisticated multivariate method may be required. A large Mahalanobis distance from an observation to the mean of all observations is one indication of a multivariate outlier (see Section 4.3).

It may be difficult to decide what to do about an outlier. If it is due to a recording error or some other definite mistake then it is fair enough to exclude it from the analysis. However, if the observation is a genuine value then this is not valid. Appropriate action then depends on the particular circumstances. Hawkins (1980) has considered the problem of outliers at some length.

11.3 Graphical methods

Graphical techniques are extremely useful for the examination of data. They should be regarded as a necessary supplement to the formal numerical calculations that have been described in this book.

It is often the graphical approach that indicates the important peculiarities of data that have a major effect on tests of significance and estimates of parameters.

An obvious problem with multivariate data is displaying values for several variables simultaneously. In recent years a number of innovative methods for doing this have been proposed. For example, Chernoff (1973) suggested making use of people's highly developed ability to detect differences in human faces. The shape of each of several features (eyes, mouth, eyebrows, etc.) can be made to correspond to different variables. Each case in a set of multivariate data generates a different face. It is then immediately apparent which cases are similar and which are different. Outliers should stand out clearly.

A useful source of more information about graphical methods for multivariate data is the book by Everitt (1978).

11.4 Missing values

Missing values can cause more problems with multivariate data than with univariate data. The trouble is that when there are many variables being measured on each individual it is quite often the case that one or two of these variables have missing values. It may then happen that if individuals with any missing values are excluded from an analysis this means excluding quite a large proportion of individuals, which may be completely impractical. For example, in studying ancient human populations skeletons are frequently broken and incomplete.

Texts on multivariate analysis are often remarkably silent on the question of missing values. To some extent this is because doing something about missing values is by no means a straightforward matter. Seber (1984) gives a discussion of the problem, with references. In practice, computer packages sometimes include a facility for estimating missing values. For example, the BMDP package (Dixon, 1983) allows missing values to be estimated by several different 'common sense' methods. One possible approach is therefore to estimate missing values and then analyse the data, including these estimates, as if they were complete data in the first place. It seems reasonable to suppose that this procedure will work satisfactorily providing that only a small proportion of values are missing.

References

Chernoff, H. (1973) Using faces to represent points in k-dimensional space graphically. *Journal of the American Statistical Association* **68**, 361–8.

Dixon, W.J. (1983) *BMDP Statistical Software*. University of California Press, Berkeley.

Everitt, B. (1978) *Graphical Techniques for Multivariate Data*. Heinemann Educational Books, London.

Hawkins, D.M. (1980) *Identification of Outliers*. Chapman and Hall, London.

Seber, G.A.F. (1984) *Multivariate Observations*. Wiley, New York.

Example sets of data

This Appendix contains three sets of data that can be used for trying out the different methods of analysis that are described in this book. In each case some possible approaches for analysing the data are suggested. However, readers are invited to develop alternatives to these.

Table A.1 Measurements, in centimetres, taken on 25 prehistoric goblets from Thailand. The measurements are defined in Fig. A.1.

Goblet	X_1	X_2	X_3	X_4	X_5	X_6
1	13	21	23	14	7	8
2	14	14	24	19	5	9
3	19	23	24	20	6	12
4	17	18	16	16	11	8
5	19	20	16	16	10	7
6	12	20	24	17	6	9
7	12	19	22	16	6	10
8	12	22	25	15	7	7
9	11	15	17	11	6	5
10	11	13	14	11	7	4
11	12	20	25	18	5	12
12	13	21	23	15	9	8
13	12	15	19	12	5	6
14	13	22	26	17	7	10
15	14	22	26	15	7	9
16	14	19	20	17	5	10
17	15	16	15	15	9	7
18	19	21	20	16	9	10
19	12	20	26	16	7	10
20	17	20	27	18	6	14
21	13	20	27	17	6	9
22	9	9	10	7	4	3
23	8	8	7	5	2	2
24	9	9	8	4	2	2
25	12	19	27	18	5	12

Data source: Professor C.F.W. Higham , University of Otago.

Data Set 1: Prehistoric goblets from Thailand

Table A.1 shows six measurements on each of 25 pottery goblets excavated from prehistoric sites in Thailand. Figure A.1 illustrates a typical goblet and shows how the measurements were made.

The main questions of interest with this set of data concern the similarities and differences between the individual goblets. Are there any obvious groupings of similar goblets? Is it possible to display the data graphically to show how the goblets are related? Are there any goblets that are particularly unusual?

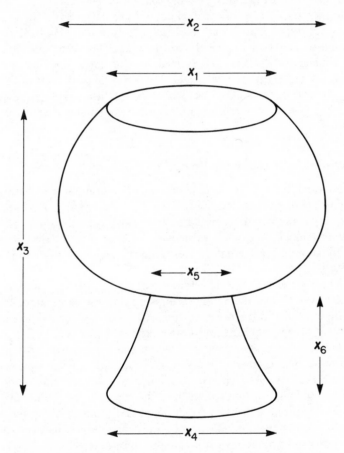

Figure A.1 Measurements made on pottery goblets from Thailand.

Possible ways to approach these questions are by cluster analysis (Chapter 8), by plotting the goblets against values for their first two principal components Chapter 5), or by carrying out a multidimensional scaling (Chapter 10). The distance matrix for a multidimensional scaling can simply be constructed of Euclidean distances between the goblets, as defined in Section 4.2.

One point that needs consideration in this example is the extent to which differences between goblets are due to shape differences rather than size differences. It may well be considered that two goblets that are almost the same shape but have very different sizes are 'similiar'. The problem of separating size and shape differences has generated a considerable number of papers and books in the scientific literature. It is not appropriate to go into this literature here. However, one simple way to try to remove the size differences between the goblets is to divide each of the measurements for a goblet by one of the measurements, say the total height of the body, or by the sum of all the measurements for that goblet. This standardization will ensure that goblets with the same shape but different sizes will have similar data values.

Data Set 2: Canine groups from Asia

Example 1.4 of Chapter 1 concerned the comparison between prehistoric dogs from Thailand and six other related animal groups in terms of mean mandible measurements. Table A.2 shows some more detailed data for the comparison of five of these groups. This is a part of the more extensive data discussed in the paper by Higham *et al.* (1980).

There are several questions that can be addressed using the data of Table A.2. Do all the animal groups display the same amount of variation in mandible measurements? Are there significant differences in mean values between the groups and, if this is so, to what extent is it possible to separate the individuals in the groups using the nine mandible measurements? Within each of the first four groups, what differences exist between males and females? Are there any outliers, i.e., individuals with measurements that appear to be anomalous?

Obvious approaches for considering these questions are by making use of various tests for significant group differences (Chapter 3) and by using discriminant function analysis (Chapter 7).

Table A.2 Values for nine mandible measurements for samples of five canine groups. The measurements are $X_1 =$ length of the mandible, $X_2 =$ breadth of mandible below 1st molar, $X_3 =$ breadth of mandible below 1st molar, $X_4 =$ height of mandible below 1st molar, $X_5 =$ length of 1st molar, $X_6 =$ breadth of 1st molar, $X_7 =$ length from 1st to 3rd molars inclusive, $X_8 =$ length from 1st to 2nd for cuon), $X_8 =$ length from 1st to 4th premolars inclusive, $X_9 =$ breadth of lower canine, all in mm.

	X_1	X_2	X_3	X_4	X_5	X_6	X_7	X_8	X_9	Sex
Modern dogs from Thailand										
1	123.4	10.1	23.1	22.8	19.2	7.8	32.2	33.0	5.6	M
2	127.4	9.6	19.0	21.9	19.2	7.8	32.2	40.4	5.8	M
3	121.1	10.2	17.8	21.0	20.6	7.9	34.5	37.5	6.2	M
4	130.1	10.7	24.1	22.4	20.1	7.9	32.4	36.8	5.9	M
5	148.6	12.0	25.0	25.2	20.8	8.4	34.7	42.7	6.6	M
6	125.0	9.5	23.3	20.3	20.4	7.8	33.3	36.9	6.3	M
7	126.2	9.1	19.7	21.5	19.2	7.5	31.8	34.7	5.5	M
8	125.2	9.7	18.8	19.2	19.1	7.5	32.3	36.7	6.2	M
9	120.8	9.6	21.8	19.6	18.5	7.6	30.5	35.1	5.3	F
10	122.2	8.9	20.3	19.9	18.9	7.6	30.5	35.2	5.7	F
11	114.5	9.3	19.0	18.9	19.9	7.8	32.6	34.0	6.5	F
12	111.5	9.1	19.4	19.8	18.7	6.6	29.8	33.0	5.1	F
13	124.4	9.3	21.3	20.7	18.1	7.1	29.7	35.5	5.5	F
14	128.1	9.6	22.4	21.0	19.0	7.5	32.2	38.0	5.8	F
15	130.4	8.4	22.9	19.7	19.0	7.3	31.1	39.9	5.8	F
16	127.3	10.5	25.0	22.5	19.5	8.7	32.2	34.9	6.1	F

Table A.2 (*Contd.*)

	X_1	X_2	X_3	X_4	X_5	X_6	X_7	X_8	X_9	Sex
Golden jackals										
1	120.0	8.2	18.1	17.3	18.1	7.0	32.0	34.7	5.2	M
2	106.8	7.9	16.6	16.5	19.7	7.0	32.1	33.6	5.3	M
3	110.0	8.1	17.6	16.0	19.0	7.1	30.7	32.5	4.7	M
4	115.6	8.5	20.0	18.3	18.0	7.1	31.5	33.1	4.7	M
5	113.6	8.2	18.7	18.1	18.8	7.9	31.5	33.0	5.1	M
6	111.0	8.5	19.0	16.4	18.1	7.1	30.0	32.7	5.0	M
7	112.9	8.5	17.3	17.6	18.5	7.1	30.0	33.6	4.6	M
8	116.5	8.7	20.2	16.7	18.3	7.0	29.9	34.3	5.2	M
9	113.6	9.4	21.0	18.5	18.7	7.5	31.1	34.8	5.3	M
10	111.9	8.2	19.0	16.8	18.5	6.8	29.7	34.0	5.1	M
11	109.5	8.5	18.3	16.9	19.2	7.0	31.1	33.2	4.9	F
12	111.3	7.7	19.9	18.2	18.0	6.7	29.7	31.9	4.5	F
13	106.9	7.2	16.5	16.0	17.5	6.0	28.0	35.0	4.7	F
14	108.0	8.2	18.4	16.2	17.5	6.5	28.7	32.5	4.8	F
15	109.5	7.3	19.2	15.5	17.4	6.1	29.8	33.3	4.5	F
16	104.6	8.3	18.6	16.9	17.2	6.5	29.2	32.4	4.5	F
17	106.9	8.4	18.0	16.9	17.7	6.2	28.6	31.0	4.3	F
18	105.5	7.8	18.9	18.4	18.1	6.2	30.6	31.6	4.4	F
19	111.2	8.4	16.6	15.9	18.2	7.0	30.3	34.0	4.7	F
20	111.0	7.6	18.7	16.5	17.8	6.5	29.9	34.8	4.6	F

Cuons

1	123.0	9.7	21.8	20.7	20.2	7.8	26.9	36.1	6.1	M
2	135.3	11.8	24.9	21.2	22.7	8.9	30.5	37.6	7.1	M
3	138.2	11.4	25.4	25.0	22.4	9.0	29.5	37.5	7.3	M
4	141.3	10.8	26.0	24.7	21.3	8.1	28.6	39.1	6.6	M
5	134.7	11.2	25.0	24.5	21.2	8.5	28.6	39.2	6.7	M
6	135.8	11.0	22.1	24.3	21.6	8.1	31.4	39.3	6.8	M
7	131.1	10.4	22.9	23.1	22.5	8.7	29.8	36.1	6.8	M
8	137.3	10.6	25.4	23.8	21.3	8.3	28.0	37.8	6.5	M
9	135.0	10.5	25.0	24.5	21.0	8.4	28.6	39.2	6.9	M
10	130.7	10.9	24.5	24.0	21.0	8.5	29.3	34.9	6.2	F
11	129.7	11.3	22.3	23.1	21.1	8.7	29.2	36.5	7.0	F
12	144.0	10.8	24.2	25.9	22.2	8.9	29.6	42.0	7.1	F
13	138.5	10.9	25.6	23.2	21.7	8.7	29.5	39.2	6.9	F
14	123.0	9.8	23.2	22.2	19.7	8.1	26.1	34.0	5.6	F
15	137.1	11.3	26.7	25.6	22.5	8.7	29.8	38.8	6.5	F
16	127.9	10.0	21.6	22.7	21.8	8.7	28.6	37.0	6.6	F
17	121.8	9.9	22.1	21.7	20.0	8.2	26.4	36.0	5.7	F

Indian wolves

1	166.8	11.5	29.0	27.9	25.3	9.5	40.5	45.2	7.2	M
2	164.3	12.3	27.0	26.0	25.3	10.0	41.6	47.3	7.9	M
3	149.5	11.5	21.4	23.5	24.6	9.3	41.3	45.5	8.5	M
4	145.5	11.3	28.0	23.8	24.3	9.2	35.5	41.2	7.2	M
5	176.8	12.4	31.3	26.6	27.3	10.5	42.9	49.8	7.9	M
6	165.8	13.4	31.7	26.5	25.5	9.5	40.3	47.0	7.3	M
7	163.6	12.1	27.1	24.3	25.0	9.9	42.1	44.5	8.3	M
8	165.1	12.6	29.5	25.5	24.7	7.7	39.9	43.4	7.9	M

Table A.2 (Contd.)

	X_1	X_2	X_3	X_4	X_5	X_6	X_7	X_8	X_9	Sex
9	131.1	11.8	19.9	23.5	22.8	8.8	37.7	40.4	6.5	F
10	163.0	10.8	27.0	24.0	24.0	9.2	39.0	47.7	7.0	F
11	164.0	10.7	24.4	23.2	25.5	9.5	42.5	47.1	7.6	F
12	140.9	10.4	19.7	22.5	22.7	8.9	38.3	42.9	6.0	F
13	148.2	10.6	26.0	21.0	23.5	8.9	38.9	39.5	7.0	F
14	157.7	10.7	25.3	24.9	24.2	9.8	40.8	45.0	7.4	F
Prehistoric Thai dogs										
1	112.0	10.1	17.0	18.2	19.0	7.7	30.7	33.1	5.8	?
2	115.0	10.0	17.9	22.5	19.5	7.8	33.0	36.0	6.0	?
3	136.0	11.9	22.4	25.0	21.2	8.5	36.2	38.7	7.0	?
4	111.1	9.9	18.7	20.1	17.5	7.3	28.5	33.5	5.3	?
5	130.4	11.2	22.5	27.4	20.0	9.1	35.2	35.3	6.6	?
6	124.9	10.7	19.0	26.1	19.5	8.4	33.3	37.0	6.3	?
7	132.5	9.6	19.2	20.2	18.7	9.7	34.8	37.9	6.6	?
8	121.0	10.7	20.5	23.0	19.3	7.9	32.0	34.9	6.0	?
9	121.5	9.8	21.8	23.0	18.5	7.9	32.2	35.2	6.1	?
10	123.5	9.5	20.1	24.3	18.7	7.6	31.5	36.5	6.0	?

Table A.3 Protein consumption (grams per head per day) in European countries.

	Red meat	White meat	Eggs	Milk	Fish	Cereals	Starchy foods	Pulses, nuts, oil-seeds	Fruits, vegetables
Albania	10.1	1.4	0.5	8.9	0.2	42.3	0.6	5.5	1.7
Austria	8.9	14.0	4.3	19.9	2.1	28.0	3.6	1.3	4.3
Belgium	13.5	9.3	4.1	17.5	4.5	26.6	5.7	2.1	4.0
Bulgaria	7.8	6.0	1.6	8.3	1.2	56.7	1.1	3.7	4.2
Czechoslovakia	9.7	11.4	2.8	12.5	2.0	34.3	5.0	1.1	4.0
Denmark	10.6	10.8	3.7	25.0	9.9	21.9	4.8	0.7	2.4
East Germany	8.4	11.6	3.7	11.1	5.4	24.6	6.5	0.8	3.6
Finland	9.5	4.9	2.7	33.7	5.8	26.3	5.1	1.0	1.4
France	18.0	9.9	3.3	19.5	5.7	28.1	4.8	2.4	6.5
Greece	10.2	3.0	2.8	17.6	5.9	41.7	2.2	7.8	6.5
Hungary	5.3	12.4	2.9	9.7	0.3	40.1	4.0	5.4	4.2
Ireland	13.9	10.0	4.7	25.8	2.2	24.0	6.2	1.6	2.9
Italy	9.0	5.1	2.9	13.7	3.4	36.8	2.1	4.3	6.7
Netherlands	9.5	13.6	3.6	23.4	2.5	22.4	4.2	1.8	3.7
Norway	9.4	4.7	2.7	23.3	9.7	23.0	4.6	1.6	2.7
Poland	6.9	10.2	2.7	19.3	3.0	36.1	5.9	2.0	6.6
Portugal	6.2	3.7	1.1	4.9	14.2	27.0	5.9	4.7	7.9
Romania	6.2	6.3	1.5	11.1	1.0	49.6	3.1	5.3	2.8
Spain	7.1	3.4	3.1	8.6	7.0	29.2	5.7	5.9	7.2
Sweden	9.9	7.8	3.5	24.7	7.5	19.5	3.7	1.4	2.0
Switzerland	13.1	10.1	3.1	23.8	2.3	25.6	2.8	2.4	4.9
UK	17.4	5.7	4.7	20.6	4.3	24.3	4.7	3.4	3.3
USSR	9.3	4.6	2.1	16.6	3.0	43.6	6.4	3.4	2.9
West Germany	11.4	12.5	4.1	18.8	3.4	18.6	5.2	1.5	3.8
Yugoslavia	4.4	5.0	1.2	9.5	0.6	55.9	3.0	5.7	3.2

Data source: Gabriel (1981) from the original by Weber (1973).

Data Set 3: Protein consumption in Europe

Table A.3 shows protein consumption in 25 European countries classified into nine food groups. For these data, possible questions of interest concern whether there are any natural groupings of the countries, whether any countries show unusual consumption patterns, and the extent to which the consumption of meat is related to the consumption of other foods. Apart from anything else, the analyses of Examples 5.2, 6.1, 7.3 and 8.1 (on employment patterns of European countries using the data of Table 1.5) can be repeated to study food consumption patterns. It would be of some interest to use canonical correlation (Chapter 9) to see how protein consumption patterns are related to employment patterns. This is possible since the countries given in Table 1.5 are almost the same as the ones given in Table A.3.

References

Gabriel, K.R. (1981) Biplot display of multivariate matrices for inspection of data and diagnosis. In *Interpreting Multivariate Data* (Ed. V. Barnett), pp. 147–73. Wiley, New York.

Higham, C.F.W., Kijngam, A. and Manly, B.F.J. (1980) An analysis of prehistoric canid remains from Thailand. *Journal of Archaeological Science* 7, 149–65.

Weber, A. (1973) *Agrarpolitik im Spannungsfeld der internationalen Ernaehrungspolitik.* Institut fuer Agrarpolitik und Marktlehre, Kiel.

Author index

Subject index